I0484388

FEDERAL EXECUTIVE TEAM

Director, Climate Change Science Program: ....William J. Brennan

Director, Climate Change Science Program Office:...............................Peter A. Schultz

Lead Agency Principal Representative to CCSP;
Deputy Under Secretary of Commerce for Oceans and Atmosphere,
National Oceanic and Atmospheric Administration:............................Mary M. Glackin

Product Lead, Chief Scientist, Physical Sciences Division, Earth
Systems Research Laboratory, National Oceanic and Atmospheric
Administration..Randall M. Dole

Synthesis and Assessment Product Coordinator,
Climate Change Science Program Office: ....Fabien J.G. Laurier

EDITORIAL AND PRODUCTION TEAM

Co-Chairs.. Randall M. Dole, NOAA
                                                                                      Martin P. Hoerling, NOAA
                                                                                      Siegfried Schubert, NASA
Federal Advisory Committee Designated Federal Official...................... Neil Christerson, NOAA
Scientific Editor .... Jessica Blunden, STG, Inc.
Scientific Editor.. Anne M. Waple, STG, Inc.
Technical Advisor.. David J. Dokken, USGCRP
Graphic Design Lead.. Sara W. Veasey, NOAA
Graphic Design Co-Lead.. Deborah B. Riddle, NOAA
Designer.. Glenn M. Hyatt, NOAA
Designer.. Deborah Misch, STG, Inc.
Designer.. Christian Zamarra, STG, Inc.
Copy Editor.. Anne Markel, STG, Inc.
Copy Editor.. Lesley Morgan, STG, Inc.
Copy Editor.. Susan Osborne, STG, Inc.
Copy Editor.. Susanne Skok, STG, Inc.
Copy Editor.. Mara Sprain, STG, Inc.
Technical Support... Jesse Enloe, STG, Inc.

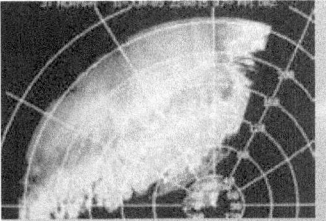

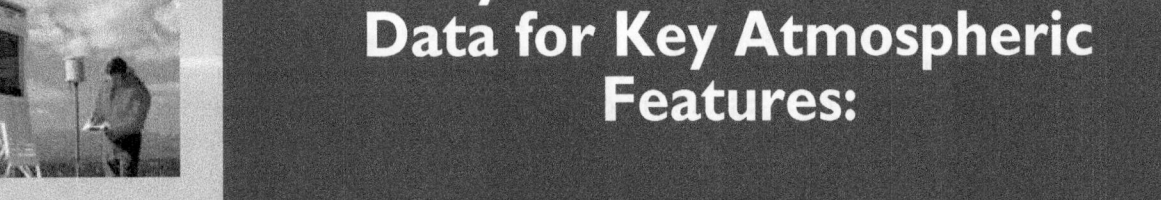

Reanalysis of Historical Climate Data for Key Atmospheric Features:

Implications for Attribution of Causes of Observed Change

Synthesis and Assessment Product 1.3
Report by the U.S. Climate Change Science Program
and the Subcommittee on Global Change Research

EDITED BY:
Randall M. Dole

December, 2008

Members of Congress:

On behalf of the National Science and Technology Council, the U.S. Climate Change Science Program (CCSP) is pleased to transmit to the President and the Congress this Synthesis and Assessment Product (SAP) *Reanalysis of Historical Climate Data for Key Atmospheric Features: Implications for Attribution of Causes of Observed Change*. This is part of a series of 21 SAPs produced by the CCSP aimed at providing current assessments of climate change science to inform public debate, policy, and operational decisions. These reports are also intended to help the CCSP develop future program research priorities.

The CCSP's guiding vision is to provide the Nation and the global community with the science-based knowledge needed to manage the risks and capture the opportunities associated with climate and related environmental changes. The SAPs are important steps toward achieving that vision and help to translate the CCSP's extensive observational and research database into informational tools that directly address key questions being asked of the research community.

This SAP addresses current capabilities to integrate observations of the climate system into a consistent description of past and current conditions through the method of reanalysis. In addition, this report assesses present capabilities to attribute causes for climate variations and trends over North America during the reanalysis period, which extends from the mid-twentieth century to the present. It was developed with broad scientific input and in accordance with the Guidelines for Producing CCSP SAPs, the Information Quality Act (Section 515 of the Treasury and General Government Appropriations Act for Fiscal Year 2001 [Public Law 106-554]), and the guidelines issued by the Department of Commerce and the National Oceanic and Atmospheric Administration pursuant to Section 515.

We commend the report's authors for both the thorough nature of their work and their adherence to an inclusive review process.

Sincerely,

Carlos M. Gutierrez
Secretary of Commerce
Chair, Committee on Climate Change
Science and Technology Integration

Samuel W. Bodman
Secretary of Energy
Vice Chair, Committee on Climate
Change Science and Technology
Integration

John H. Marburger III
Director, Office of Science and
Technology Policy
Executive Director, Committee
on Climate Change Science and
Technology Integration

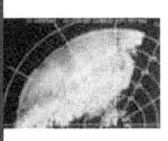

TABLE OF CONTENTS

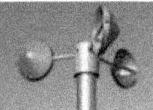

CHAPTER

1 5

Introduction

2 11

Reanalysis of Historical Climate Data for Key Atmospheric Features

TABLE OF CONTENTS

TABLE OF CONTENTS

AUTHOR TEAM FOR THIS REPORT

Preface   **Convening Lead Author:** Randall Dole, NOAA

Executive Summary   **Convening Lead Author:** Randall Dole, NOAA
                         Lead Authors: Martin Hoerling, NOAA; Siegfried Schubert, NASA

Chapter 1   **Convening Lead Author:** Randall Dole, NOAA
                         Lead Author: Martin Hoerling, NOAA

Chapter 2   **Convening Lead Author:** Siegfried Schubert, NASA
                         Lead Authors: Phil Arkin, University of Maryland; James Carton,
                         University of Maryland; Eugenia Kalnay, University of Maryland;
                         Randal Koster, NASA
                         Contributing Authors: Randall Dole, NOAA; Roger Pulwarty,
                         NOAA

Chapter 3   **Convening Lead Author:** Martin Hoerling, NOAA
                         Lead Authors: Gabriele Hegerl, University of Edinburgh; David
                         Karoly, University of Melbourne; Arun Kumar, NOAA; David Rind,
                         NASA
                         Contributing Author: Randall Dole, NOAA

Chapter 4   **Convening Lead Author:** Randall Dole, NOAA
                         Lead Authors: Martin Hoerling, NOAA; Siegfried Schubert, NASA

Appendix A   **Convening Lead Authors:** James Carton, University of Maryland;
                         Eugenia Kalnay, University of Maryland

Appendix B   **Convening Lead Author:** Martin Hoerling, NOAA
                         Lead Authors: Gabriele Hegerl, University of Edinburgh; David
                         Karoly, University of Melbourne; Arun Kumar, NOAA; David Rind,
                         NASA

ACKNOWLEDGEMENT

CCSP Synthesis and Assessment Product 1.3 (SAP 1.3) was developed with the benefit of a scientifically rigorous, first draft peer review conducted by a committee appointed by the National Research Council (NRC). Prior to their delivery to the SAP 1.3 Author Team, the NRC review comments, in turn, were reviewed in draft form by a second group of highly qualified experts to ensure that the review met NRC standards. The resultant NRC Review Report was instrumental in shaping the final version of SAP 1.3, and in improving its completeness, sharpening its focus, communicating its conclusions and recommendations, and improving its general readability.

We wish to thank the members of the NRC Review Committee: David H. Bromwich (Chair), The Ohio State University; Aiguo Dai, National Center for Atmospheric Research; Ioana M. Dima, AIR Worldwide Corporation; John W. Nielsen-Gammon, Texas A&M University; Benjamin Kirtman, University of Miami; Robert N. Miller, Oregon State University; and Andrew W. Robertson, International Research Institute for Climate and Society. The authors also thank the NRC Staff members who coordinated the process: Chris Elfring, Director, Board on Atmospheric Sciences and Climate; Maria Uhle, Study Director; Rachael Shiflett, Senior Program Assistant; and Shuba Banskota, Financial Associate.

We also thank the individuals who reviewed the NRC Report in its draft form: Mary Anne Carroll, University of Michigan; Peter R. Leavitt, Weather Information Company; Elizabeth L. Malone, Joint Global Change Research Institute; Joellen L. Russell, University of Arizona; and Andrew R. Solow, Woods Hole Oceanographic Institution.

The review process for SAP 1.3 also included a public review of the Second Draft, and we thank the individuals who participated in this important step. The Author Team carefully considered all comments, and the subsequent revisions resulted in further improvements in the quality and clarity of the report. Note that the respective review bodies were not asked to endorse the final version of SAP 1.3, as this is the responsibility of the National Science and Technology Council.

The authors would like to give special thanks to Jon Eischeid of the NOAA Earth System Research Laboratory and Sara Veasey of NOAA's National Climatic Data Center for their great help in preparing numerous figures for publication, and to Jessica Blunden of STG, Inc., for significantly clarifying and improving the report in her role as Scientific Editor. We wish to thank Glenn Hyatt and Deborah Riddle of NOAA's National Climatic Data Center, and Deb Misch, Susan Osborne, Susanne Skok, Mara Sprain, and Christian Zamarra for their dedication to the production of SAP 1.3. The authors also thank Neil Christerson and Krisa Arzayus of the NOAA Climate Program Office for their able support in shepherding this Product through the many steps required for its development.

ABSTRACT

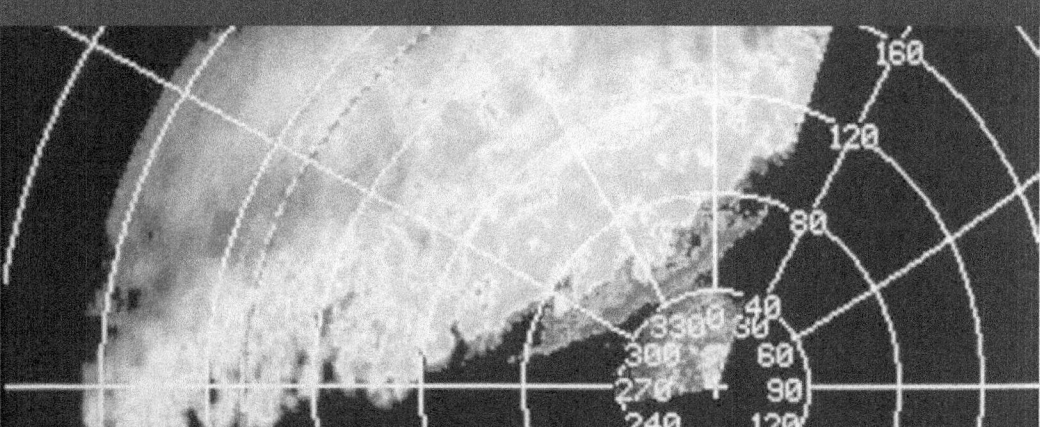

This Climate Change Science Program Synthesis and Assessment Product addresses current capabilities to integrate observations of the climate system into a consistent description of past and current conditions through the method of reanalysis. In addition, the Product assesses present capabilities to attribute causes for climate variations and trends over North America during the reanalysis period, which extends from the mid-twentieth century to the present.

This Product reviews the strengths and limitations of current atmospheric reanalysis products. It finds that reanalysis data play a crucial role in helping to identify, describe, and understand atmospheric features associated with weather and climate variability, including high-impact events such as major droughts and floods. Reanalysis data play an important role in assessing the ability of climate models to simulate the average climate and its variations. The data also help in identifying deficiencies in representations of physical processes that produce climate model errors.

The Product emphasizes that significant improvements are possible that would substantially increase the value of reanalyses for climate research, applications, and decision support. Advances are likely through developing new methods to address changes in observing systems over time, improving the historical observational database, and developing integrated Earth system models and analysis systems that include key climate elements for decision support that were not contained in initial atmospheric reanalyses, such as a carbon cycle, aerosols and other important atmospheric constituents.

The Product also assesses current understanding of the causes of observed North American climate variability and trends from 1951 to 2006. This assessment is based on results from research studies, climate model simulations, and reanalysis and observational data. For annual, area-average surface temperatures over North America, more than half of the observed surface warming since 1951 is likely due to anthropogenic forcing associated with greenhouse gas forcing. However, warming due to anthropogenic greenhouse gas emissions alone is unlikely to be the main cause for regional and seasonal differences of surface temperature changes, such as the absence of a summertime warming trend over the Great Plains of the United States and the absence of a warming trend in both winter and summer over portions of the southern United States.

The regional and seasonal variations in temperature trends are related to the principal atmospheric wind patterns that affect North American climate, which are well represented in climate reanalyses. It is likely that variations in regional sea surface temperatures have played an important role in forcing these atmospheric wind patterns, although there is evidence that some wind changes are also due to anthropogenic forcing.

In contrast to temperature, there is no discernible trend during this period in annual average North American precipitation, although there is substantial interannual-to-decadal variability. Part of the observed variability in precipitation appears to be related to regional variations of sea surface temperatures during this period.

RECOMMENDED CITATIONS

Entire Report:

CCSP, 2008: *Reanalysis of Historical Climate Data for Key Atmospheric Features: Implications for Attribution of Causes of Observed Change.* A Report by the U.S. Climate Change Science Program and the Subcommittee on Global Change Research [Randall Dole, Martin Hoerling, and Siegfried Schubert (eds.)]. National Oceanic and Atmospheric Administration, National Climatic Data Center, Asheville, NC, 156 pp.

Preface:

Dole, R., 2008: Preface. In: *Reanalysis of Historical Climate Data for Key Atmospheric Features: Implications for Attribution of Causes of Observed Change.* A Report by the U.S. Climate Change Science Program and the Subcommittee on Global Change Research [Randall Dole, Martin Hoerling, and Siegfried Schubert (eds.)]. National Oceanic and Atmospheric Administration, National Climatic Data Center, Asheville, NC, pp. XI–XIV

Executive Summary:

Dole, R., M. Hoerling, and S. Schubert, 2008: Executive summary. In: *Reanalysis of Historical Climate Data for Key Atmospheric Features: Implications for Attribution of Causes of Observed Change.* A Report by the U.S. Climate Change Science Program and the Subcommittee on Global Change Research [Randall Dole, Martin Hoerling, and Siegfried Schubert (eds.)]. National Oceanic and Atmospheric Administration, National Climatic Data Center, Asheville, NC, pp. 1–4

Chapter 1:

Dole, R. and M. Hoerling, 2008: Introduction. In: *Reanalysis of Historical Climate Data for Key Atmospheric Features: Implications for Attribution of Causes of Observed Change.* A Report by the U.S. Climate Change Science Program and the Subcommittee on Global Change Research [Randall Dole, Martin Hoerling, and Siegfried Schubert (eds.)]. National Oceanic and Atmospheric Administration, National Climatic Data Center, Asheville, NC, pp. 5–10

Chapter 2:

Schubert, S., P. Arkin, J. Carton, E. Kalnay, and R. Koster, 2008: Reanalysis of historical climate data for key atmospheric features. In: *Reanalysis of Historical Climate Data for Key Atmospheric Features: Implications for Attribution of Causes of Observed Change.* A Report by the U.S. Climate Change Science Program and the Subcommittee on Global Change Research [Randall Dole, Martin Hoerling, and Siegfried Schubert (eds.)]. National Oceanic and Atmospheric Administration, National Climatic Data Center, Asheville, NC, pp. 11–46

Chapter 3:

Hoerling, M., G. Hegerl, D. Karoly, A. Kumar, and D. Rind, 2008: Attribution of the causes of climate variations and trends over North America during the modern reanalysis period. In: *Reanalysis of Historical Climate Data for Key Atmospheric Features: Implications for Attribution of Causes of Observed Change.* A Report by the U.S. Climate Change Science Program and the Subcommittee on Global Change Research [Randall Dole, Martin Hoerling, and Siegfried Schubert (eds.)]. National Oceanic and Atmospheric Administration, National Climatic Data Center, Asheville, NC, pp. 47–92

Chapter 4:

Dole, R., M. Hoerling, and S. Schubert, 2008: Recommendations. In: *Reanalysis of Historical Climate Data for Key Atmospheric Features: Implications for Attribution of Causes of Observed Change.* A Report by the U.S. Climate Change Science Program and the Subcommittee on Global Change Research [Randall Dole, Martin Hoerling, and Siegfried Schubert (eds.)]. National Oceanic and Atmospheric Administration, National Climatic Data Center, Asheville, NC, pp. 93–104

Appendix A:

Carton, J. and E. Kalnay, 2008: Appendix A: Data assimilation. In: *Reanalysis of Historical Climate Data for Key Atmospheric Features: Implications for Attribution of Causes of Observed Change.* A Report by the U.S. Climate Change Science Program and the Subcommittee on Global Change Research [Randall Dole, Martin Hoerling, and Siegfried Schubert (eds.)]. National Oceanic and Atmospheric Administration, National Climatic Data Center, Asheville, NC, pp. 105–106

Appendix B:

Hoerling, M., G. Hegerl, D. Karoly, A. Kumar, and D. Rind, 2008: Appendix B: Data and methods used for attribution. In: *Reanalysis of Historical Climate Data for Key Atmospheric Features: Implications for Attribution of Causes of Observed Change.* A Report by the U.S. Climate Change Science Program and the Subcommittee on Global Change Research [Randall Dole, Martin Hoerling, and Siegfried Schubert (eds.)]. National Oceanic and Atmospheric Administration, National Climatic Data Center, Asheville, NC, pp. 107–114

PREFACE

Report Motivation and Guidance for Using this Synthesis/Assessment Product

Convening Lead Author: Randall Dole, NOAA/ESRL

A primary objective of the U.S. Climate Change Science Program (CCSP) is to provide the best possible scientific information to support public discussion, and government and private sector decision making on key climate-related issues. To help meet this objective, the CCSP has identified 21 Synthesis and Assessment Products (SAPs) that address its highest priority research, observational, and decision-support needs. This Product, CCSP SAP 1.3, is one of three products developed to address the first goal of the CCSP Strategic Plan: Improve knowledge of the Earth's past and present climate and environment, including its natural variability, and improve understanding of the causes of observed variability and change. This Product assesses present capabilities to describe key features of climate from the mid-twentieth century to the present through the scientific method of reanalysis. It also assesses current understanding of the causes of observed climate variability and changes over the North American region during this same period.

P.1 OVERVIEW OF PRODUCT

New climate observations are most informative when they can be put in the context of what has occurred in the past. Are current conditions unusual or have they been observed frequently before? Are the current conditions part of a long-term trend or a manifestation of climate variability that may be expected to reverse over months, seasons, or years? Are similar or related changes occurring in other parts of the globe? What are the processes and mechanisms that can explain current conditions, and how are they similar to, or different from, what has occurred in the past?

The scientific methods of climate reanalysis and attribution are central to addressing such questions. In brief, reanalysis is a method for constructing a high-quality record of past climate conditions. Attribution is the process of establishing the most likely cause (or causes) for an observed climate variation or change.

An important goal of the reanalysis efforts assessed in this Product is to provide comprehensive, consistent, and reliable long-term datasets of temperatures, precipitation, winds, and numerous other variables that characterize the state of the climate system. Because these datasets provide continuous time records, typically at six-hour intervals over several decades, they play an important role in documenting how weather and climate conditions are changing over time. The comprehensive nature of climate reanalyses also makes such datasets of great value in helping scientists to better understand the often complex relationships among variables, for example, how changes in temperatures may be connected to changes in winds, and how these in turn may be related to changes in cloudiness and precipitation.

Reanalysis datasets provide a foundation for a broad range of weather and climate research. As one measure of their extraordinary research impact, an overview paper describing one of the initial reanalyses produced in the United States is now the most widely cited paper in the geophysical sciences. Beyond their research applications, products derived from reanalysis data are used in an increasing range of commercial and business applications in sectors such as energy, agriculture, water resources, and insurance. Some commonly used products include maps showing monthly and seasonal averages, variability and trends in temperatures, winds, precipitation and storminess.

Increasingly, climate scientists are also being asked to go beyond descriptions of *what* are the current climate conditions and how they compare with the past to also explain *why* climate is evolving as observed; that is, to provide attribution for the causes of observed climate variations and change. The capability to attribute causes for past and current climate conditions is an important factor in developing public confidence in scientific understanding of mechanisms that produce climate variability and change. Attribution also provides a scientific underpinning for predicting future climate as well as information useful for evaluating needs and options for adaptation and/or mitigation.

This Product addresses the strengths and limitations of current reanalysis products in documenting, integrating, and advancing knowledge of the climate system. It also assesses present scientific capabilities to attribute causes for weather and climate variations and trends over North America during the reanalysis period (from the mid-twentieth century to the present), including the uses, limitations, and opportunities for improvement of reanalysis data applied for this purpose.

The Product is intended to be of value to the following users:

- policymakers in assessing current scientific capabilities to attribute causes of climate variations and change over the North American region;
- scientists and other users of reanalysis data through the assessment of strengths and limitations of current reanalyses; and
- science program managers in developing priorities for future observing, modeling, and analysis systems required to advance national and international capabilities in climate reanalysis and attribution.

Following guidance provided by the Climate Change Science Program, this Product is written primarily for the informed lay reader. For subject matter experts, more detailed discussions are available through the original references cited herein. Because some terms will be new to non-specialists, a glossary and a list of acronyms and abbreviations are included at the end of this Product.

P.2 PRIMARY FOCUS OF THE PRODUCT

Chapter 1 provides a brief, non-technical discussion of the fundamental concepts of reanalysis and attribution. Two issues of broad interest follow, within which specific questions are addressed: (1) the reanalysis of historical climate data for key atmospheric features, in particular, for past climate variations and trends over the reanalysis period from the mid-twentieth century to the present, and (2) attribution of the causes of climate variations and trends over North

America during the same period. These topics are described in more detail below.

P.2.1 Reanalysis of Historical Climate Data for Key Atmospheric Features
The availability and usefulness of reanalysis data have led to many important scientific advances as well as a broad range of new applications. However, limitations of past and current observations, models, and reanalysis methods have each contributed to uncertainties in describing climate system behavior. Chapter 2 focuses on the strengths and limitations of current reanalysis data for identifying and describing past climate variations and trends.

The first global atmospheric reanalyses were developed a little over a decade ago by NASA, NOAA (together with the National Center for Atmospheric Research [NCAR]), and the European Centre for Medium Range Weather Forecasts. These initial reanalyses were constructed by combining observations from diverse data sources within sophisticated models used for weather predictions through a process called data assimilation. Because of the origins in the use of weather models, the initial reanalyses and the majority of those conducted since that time have focused on reconstructing past atmospheric conditions. The longest reanalysis, conducted by NOAA and NCAR, extends back to 1948. Because of their maturity and extensive use, atmospheric reanalyses constitute the primary focus of this Product. However, efforts are now underway to create reanalyses for other components of the Earth's climate system, such as the ocean and land surface; emerging capabilities in these areas will also be briefly discussed.

The key questions addressed in Chapter 2 are:
- What is a climate reanalysis? What role does reanalysis play within a comprehensive climate observing system?
- What can reanalysis tell us about climate processes and their representation in models used for climate predictions and climate change projections?
- What is the capacity of current reanalyses to help identify and understand major seasonal-to-decadal climate variations, including changes in the frequency and intensity of climate extremes such as droughts?
- To what extent is there agreement or disagreement between climate trends in surface temperature and precipitation derived from reanalyses and those derived from independent data; that is, from data that are not included in constructing the reanalysis?
- What steps would be most useful in reducing false jumps and trends in climate time series (those that may be due to changes in observing systems or other non-physical causes) and other uncertainties in past climate conditions? Specifically, what contributions could be made

through advances in data recovery or quality control, modeling, and/or data assimilation techniques?

The assessment of capabilities and limitations of current reanalysis datasets for various purposes will be of value for determining best uses of current reanalysis products for scientific and practical purposes. This Chapter will also be useful for science program managers in developing priorities for improving the scientific and practical value of future climate reanalyses.

P.2.2 Attribution of the Causes of Climate Variations and Trends over North America

Chapter 3 discusses current understandings of the causes of climate variations and trends over North America from the mid-twentieth century to the present, the time period encompassed by current atmospheric reanalysis products. It also addresses strengths and limitations of reanalysis products in supporting research to attribute the causes of climate variations and trends over North America during this time period. The key questions are:

- What is climate attribution? What are the scientific methods used for establishing attribution?
- What is the present understanding of the causes for North American climate trends in annual temperature and precipitation during the reanalysis record?
- What is the present understanding of causes for seasonal and regional variations in U.S. temperature and precipitation trends over the reanalysis record?
- What are the nature and causes of apparent rapid climate shifts relevant to North America over the reanalysis record?
- What is the present understanding of the causes for high-impact drought events over North America during the reanalysis record?

This Chapter will provide policy makers with an assessment of current scientific understanding and remaining uncertainties regarding the causes of major climate variations and trends over North America since the mid-twentieth century. Resource managers and other decision makers, as well as the general public, will also benefit from this assessment.

Finally, Chapter 4 discusses steps needed to improve national capabilities in reanalysis and attribution to better address key questions in climate science and to increase the value of future reanalysis and attribution products for applications and decision making. This Chapter will be of value to scientists and research program managers who are engaged in efforts to advance national and international capabilities in climate reanalysis and attribution.

P.3 TREATMENT OF UNCERTAINTY

Terms used in this Product to indicate the assessed likelihood of an outcome are consistent with those used in the Intergovernmental Panel on Climate Change (IPCC) Fourth Assessment Report (*Climate Change 2007: The Physical Science Basis*) and summarized in Table P.1.

Terms denoting levels of confidence in findings are also consistent with the IPCC Fourth Assessment Report usage, as specified in Table P.2.

Table P.1 Terminology regarding likelihood of outcome according to IPCC AR4.

Likelihood Terminology	Likelihood of occurrence/outcome
Virtually Certain	more than 99 percent probability
Extremely Likely	more than 95 percent probability
Very Likely	more than 90 percent probability
Likely	more than 66 percent probability
More Likely than Not	more than 50 percent probability
About as Likely as Not	33 to 66 percent probability
Unlikely	less than 33 percent probability
Very Unlikely	less than 10 percent probability
Extremely Unlikely	less than 5 percent probabillity
Exceptionally Unlikely	less than 1 percent probability

Table P.2 Terminology regarding degree of confidence according to IPCC AR4.

Terminology	Degree of confidence in being correct
Very High Confidence	At least nine out of ten chance of being correct
High Confidence	About eight out of ten chance
Medium Confidence	About five out of ten chance
Low Confidence	About two out of ten chance
Very Low Confidence	Less than one out of ten chance

P.4 SCOPE AND LIMITATIONS OF THIS PRODUCT

The time period considered in this Product is limited to that of present-day reanalysis datasets, which extend from 1948 to the present. As discussed in Chapter 4, an effort is now underway to extend reanalysis data back to at least the latter part of the nineteenth century. While initial results appear promising, this extended reanalysis project is not yet complete; therefore, it is not possible to assess the preliminary results in this Product.

The findings presented in this Product provide a snapshot of the current state of knowledge as of mid-2007. The fields of climate analysis, reanalysis, and attribution are cutting edge areas of climate research, with new results being obtained every month. Within the next few years new results are likely to appear that will supersede some of the key findings discussed in this Product; for example, with respect to the quality, types, and lengths of reanalysis records now available.

The scope of this Product was considered in light of other ongoing assessments, in particular the IPCC Fourth Assessment Report and other synthesis and assessment reports being developed within the Climate Change Science Program. The IPCC Report emphasizes climate change at global to continental scales. This Product focuses on the United States/North American sector and considers regional climate variations and trends of specific interest to U.S. resource managers, decision makers, and the general public.

EXECUTIVE SUMMARY

Convening Lead Author: Randall Dole, NOAA/ESRL

Lead Authors: Martin Hoerling, NOAA/ESRL; Siegfried Schubert, NASA/GMAO

Among the most common questions that climate scientists are asked to address are: What are current climate conditions? How do these conditions compare with the past? What are the causes for current conditions, and are the causes similar to or different from those of the past? This Climate Change Science Program (CCSP) Synthesis and Assessment Product considers such questions, focusing on advances in scientific understanding obtained through the methods of reanalysis and attribution.

In climate science, a *reanalysis* is a method for constructing a high-quality climate record that combines a diverse array of past observations together within a model to derive a best estimate of how the climate system has evolved over time. An important goal of the reanalysis efforts assessed in this Product is to provide comprehensive, consistent, and reliable long-term datasets of temperatures, precipitation, winds, and numerous other variables that characterize the state of the climate system. The atmospheric reanalyses assessed in this Product provide a continuous, detailed record of how the atmosphere has evolved every 6 to 12 hours over periods spanning multiple decades. The Product addresses the strengths and limitations of current reanalyses in advancing scientific knowledge of the climate system. It then assesses current scientific capabilities to attribute causes for climate variations and trends over North America during the reanalysis period, which extends from the mid-twentieth century to the present. The Product concludes with recommendations to improve national capabilities in reanalysis and attribution in order to increase the value of future products for research, applications and decision making.

This Product represents a significant extension beyond the recently completed Intergovernmental Panel on Climate Change (IPCC) Fourth Assessment Report (*Climate Change 2007: The Physical Science Basis*). While the IPCC Report mainly emphasized climate change at global to continental scales, this Product focuses on North America, including regional climate variations and trends that are of substantial interest to the U.S. general public, decision makers, and policy makers.

ES.I PRIMARY RESULTS AND FINDINGS

ES.1.1 Strengths and Limitations of Current Reanalysis Datasets for Representing Key Atmospheric Features

KEY FINDINGS
(from Chapter 2)

- Reanalysis plays a crucial integrating role within a global climate observing system by producing comprehensive, long-term, objective, and consistent records of climate system components, including the atmosphere, oceans and land surface.

- Reanalysis data play a fundamental and unique role in studies that address the nature, causes, and impacts of global-scale and regional-scale climate phenomena.

- Reanalysis datasets are of great value in studies of the physical processes that produce high-impact weather and climate events such as droughts and floods, as well as other key atmospheric features that affect the United States, including climate variations associated with major modes of climate variability, such as the El Niño-Southern Oscillation.

- Global and regional surface temperature trends in reanalysis datasets are broadly consistent with those obtained from temperature datasets constructed from surface observations not included in the reanalyses, particularly since the late 1970s. However, in some regions (*e.g.*, Australia) the reanalysis trends show major differences with observations.

- Reanalysis precipitation trends are less consistent with those calculated from observational datasets. The differences are likely due principally to limitations in the initial reanalysis models and the methods used for integrating diverse datasets within models.

- Current reanalysis data are extremely valuable for a host of scientific and practical applications; however, the overall quality of reanalysis products varies with latitude, altitude, time period, location and time scale, and variable of interest, such as temperature, winds or precipitation.

- Current global reanalysis data are most reliable in Northern Hemisphere midlatitudes, in the middle to upper troposphere (about three to twelve miles above Earth's surface), and for regional and larger areas. They are also most reliable for time periods ranging from one day up to several years, making reanalysis data well suited for studies of midlatitude storms and short-term climate variability.

- Present reanalyses are more limited in their value for detecting long-term climate trends, although there are cases where reanalyses have been usefully applied for this purpose. Important factors constraining the value of reanalyses for trend detection include changes in observing systems over time; deficiencies in observational data quality and spatial coverage; model limitations in representing interactions across the land-atmosphere and ocean-atmosphere interfaces, which affect the quality of surface and near-surface weather and climate variables; and inadequate representation of the water cycle.

- At the present time, datasets constructed for an individual variable, for example, surface temperature or precipitation, are generally superior for climate change detection. However, the integrated and comprehensive nature of reanalysis data provides a quantitative foundation for improving understanding of the processes that produce changes. These qualities make reanalysis data more useful than individual variable data sets for attributing the causes of climate variations and change.

ES.1.2 Attribution of the Causes of Climate Variations and Trends over North America during the Modern Reanalysis Period

KEY FINDINGS
(from Chapter 3)

- Significant advances have occurred over the past decade in capabilities to attribute causes for observed climate variations and change.

- Methods now exist for establishing attribution for the causes of North American climate variations and trends due to internal climate variations and/or changes in external climate forcing.

Annual, area-average change for the period 1951 to 2006 across North America shows the following:

- Seven of the warmest ten years for annual surface temperatures from 1951 to 2006 have occurred between 1997 and 2006.

- The 56-year linear trend (1951 to 2006) of annual surface temperature is +0.90°C ±0.1°C (1.6°F ± 0.2°F).

- Virtually all of the warming since 1951 has occurred after 1970.

- More than half of this warming is *likely* the result of human-caused greenhouse gas forcing of climate change.

- Changes in ocean temperatures *likely* explain a substantial fraction of the human-caused warming of North America.

- There is no discernible trend in average precipitation since 1951, in contrast to trends observed in extreme precipitation events.

Spatial variations in annual average change for the period from 1951 to 2006 across North America show the following:

- Observed surface temperature change has been largest over northern and western North America, with up to +2°C (3.6°F) warming in 56 years over Alaska, the Yukon Territories, Alberta, and Saskatchewan.

- Observed surface temperature change has been smallest over the southern United States and eastern Canada, where no significant trends have occurred.

- There is *very high* confidence that changes in atmospheric wind patterns have occurred, based upon reanalysis data, and that these wind pattern changes are *likely* the physical basis for much of the spatial variations in surface temperature change over North America, especially during winter.

- The spatial variations in surface temperature change over North America are *unlikely* to be the result of anthropogenic forcing alone.

- The spatial variations in surface temperature change over North America are *very likely* influenced by changes in regional patterns of sea surface temperatures through the effects of sea surface temperatures on atmospheric wind patterns, especially during winter.

Spatial variations of seasonal average change for the period 1951 to 2006 across the United States show that:

- Six of the warmest ten summers and winters for the contiguous United States average surface temperatures from 1951 to 2006 occurred recently (1997 to 2006).

- During summer, surface temperatures warmed most over western states, with insignificant change between the Rocky and Appalachian Mountains. During winter, surface temperatures warmed most over northern and western states, with insignificant changes over Maine and the central Gulf of Mexico.

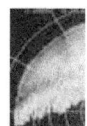

- The spatial variations in summertime surface temperature change are *unlikely* to be the result of anthropogenic forcing alone.

- The spatial variations and seasonal differences in precipitation change are *unlikely* to be the result of anthropogenic greenhouse gas forcing alone.

- Some of the spatial variations and seasonal differences in precipitation change and variations are *likely* the result of regional variations in sea surface temperatures.

An assessment to identify and attribute the causes of abrupt climate change over North America for the period 1951 to 2006 finds that:

- There are limitations for detecting rapid climate shifts and distinguishing these shifts from quasi-cyclical variations because current reanalysis data only extends back until to the mid-twentieth century. Reanalysis over a longer time period is needed to distinguish between these possibilities with scientific confidence.

An assessment to determine trends and attribute causes for droughts for the period 1951 to 2006 shows that:

- It is *unlikely* that a systematic change has occurred in either the frequency or area coverage of severe drought over the contiguous United States from the mid-twentieth century to the present.

- It is *very likely* that short-term (monthly-to-seasonal) severe droughts that have impacted North America during the past half-century are mostly due to atmospheric variability, in some cases amplified by local soil moisture conditions.

- It is *likely* that sea surface temperature variations have been important in forcing long-term (multi-year) severe droughts that have impacted North America during the past half-century.

- It is *likely* that anthropogenic warming has increased drought impacts over North America in recent decades through increased water stresses associated with warmer conditions, but the magnitude of the effect is uncertain.

ES.2 RECOMMENDATIONS

The following six recommendations are aimed at improving the scientific and practical value of future reanalyses of the climate system.

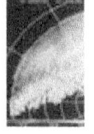

- To better detect changes in the climate system, improve the quality and consistency of the observational data and reduce effects of observing system changes.

- Develop analysis methods that are optimized for climate research and applications. These methods should include uncertainty estimates for all reanalysis products.

- To improve the description and understanding of major climate variations that occurred prior to the mid-twentieth century, develop the longest possible consistent record of past climate conditions.

- To improve decision support, develop future climate reanalysis products at finer space scales (*e.g.*, resolutions of 10 miles rather than 100 miles) and emphasize products that are most relevant for applications, such as surface temperatures, winds, cloudiness, and precipitation.

- Develop new national capabilities in analysis and reanalysis that focus on variables that are of high relevance to policy and decision support. Such variables include those required to monitor changes in the carbon cycle and to understand interactions among Earth system components (atmosphere, ocean, land, cryosphere, and biosphere) that may lead to accelerated or diminished rates of climate change.

- Develop a more coordinated, effective, and sustained national capability in analysis and reanalysis to support climate research and applications.

The following priorities are recommended for reducing uncertainties in climate attribution and realizing the benefits of this information for decision support:

- Develop a national capability in climate attribution to provide regular and reliable explanations of evolving climate conditions relevant to decision making.

- Focus research to better explain causes of climate conditions at regional and local levels, including the roles of changes in land cover, land use, atmospheric aerosols, greenhouse gases, sea surface temperatures, and other factors that contribute to climate change.

- Explore a range of methods to better quantify and communicate findings from attribution research.

CHAPTER 1

Introduction

Convening Lead Author: Randall Dole, NOAA/ESRL

Lead Author: Martin Hoerling, NOAA/ESRL

FUNDAMENTAL CONCEPTS

Among the most frequent questions that the public and decision makers ask climate scientists are: What do we know about past climate? What are the uncertainties in observations of climate? What do we know about the causes of climate variations and change? What are the uncertainties in explaining the causes for observed climate conditions? The scientific methods of climate *reanalysis* and *attribution* play important roles in helping to address such questions. This Chapter is intended to provide readers with an initial foundation for understanding the nature and scientific roles of reanalysis and attribution, as well as their potential relevance for applications and decision making. These subjects are then discussed in detail in the remainder of the Product.

1.1 REANALYSIS

In atmospheric science, an *analysis* is a detailed representation of the state of the atmosphere that is based on observations (Geer, 1996). More generally, an analysis may also be performed for other parts of the climate system, such as the oceans or land surface. The analysis is often displayed as a map depicting the values of a single variable such as air temperature, wind speed, or precipitation amount, or of multiple variables for a specific time period, level, and region. The daily weather maps that are presented in newspapers, on television, and in numerous other sources are familiar examples of this form of analysis (Figure 1.1a). Analyses are also performed at levels above the Earth's surface (Figure 1.1b) in order to provide a complete depiction of atmospheric conditions throughout the depth of the atmosphere. This type of analysis enables atmospheric scientists to locate key atmospheric features, such as the jet stream, and plays a crucial role in weather forecasting by providing initial conditions required for models used for weather prediction.

A retrospective analysis, or *reanalysis*, is an objective, quantitative method for producing a high quality sequence of analyses that extends over a sufficiently long time period to have value for climate applications (as well as for other purposes). An important goal of most reanalysis efforts to date has been to provide an accurate and consistent long-term data record of the global atmosphere. As discussed in Chapter 2, reanalyses have also been conducted or are in progress for the oceans and land surface. In certain cases, a reanalysis may be performed for a single variable, such as precipitation or surface temperature (Fuchs, 2007). However, in many modern atmospheric reanalyses the goal is to develop an accurate and physically consistent representation of an extensive set of variables (*e.g.*, winds, temperatures, pressures, *etc.*) required to provide a comprehensive, detailed depiction of how the atmosphere has evolved over an extended period of time (typically, decades). Such comprehensive reanalyses are a major focus of this assessment.

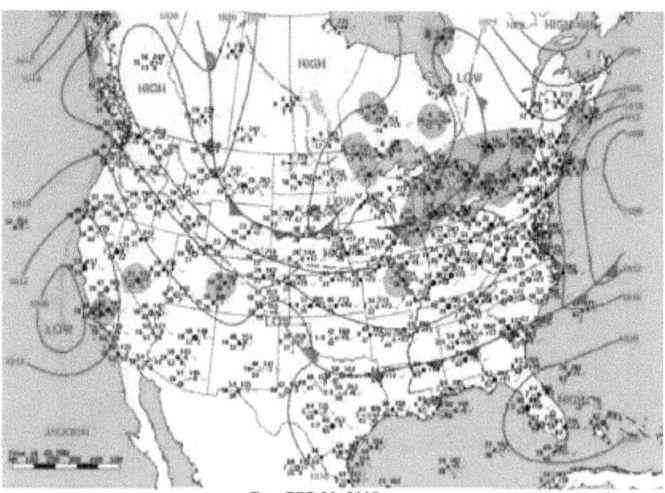

Tue, FEB 22, 2005
Surface Weather Map and Station Weather at 7:00 A.M. E.S.T.

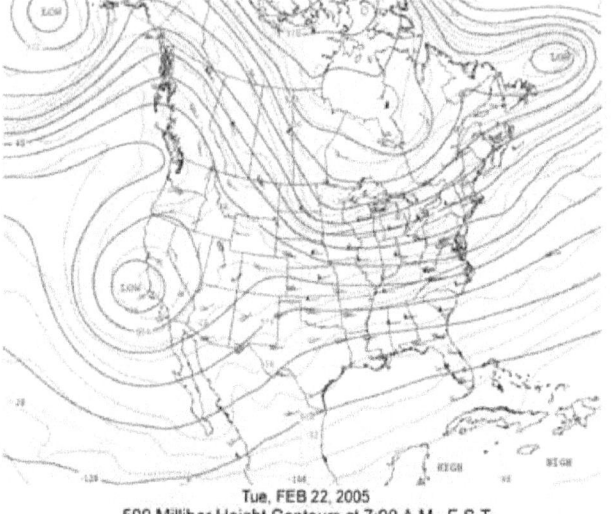

Tue, FEB 22, 2005
500 Millibar Height Contours at 7:00 A.M. E.S.T.

Figure 1.1 Examples of map analyses for a given day (February 22, 2005) for the continental United States and adjacent regions. (a) Surface weather analysis, or "weather map". Contours are lines of constant pressure (isobars), while green shaded areas denote precipitation. Positions of low pressure and high pressure centers, fronts and a subset of surface station locations providing observations that underpin the analysis are also shown. (b) A map of the heights (solid lines, in decameters) and temperatures (dotted lines, in °C) of a constant pressure surface, in this case the 500 millibar surface, which represents conditions at an elevation of approximately 18,000 feet. The symbols with bars and/or pennants show wind speeds and directions obtained from observations. Wind directions "blow" from the end with bars toward the open end, the open end depicting the observation station location (e.g., winds over Denver, Colorado on this day are from the west, while those over Oakland, California are from the east). Note the strong relationship between the wind direction and the height contours, with the station winds blowing nearly parallel to the height contour lines shown in the analysis (and counter-clockwise around lows, as for example the low center just off the California coast). This is an example of a balanced relationship that is used to help construct the analyses, as discussed in Chapter 2.

The reanalysis efforts assessed in this Product estimate past conditions using a method that integrates observations from numerous data sources (Figure 1.2) together within a state-of-the art atmospheric model (or a model of another climate system component, such as the ocean or land surface). This data-model integration provides a comprehensive, high quality, temporally continuous, and physically consistent dataset of atmospheric variables for use in climate research and applications. The models provide physical consistency by constraining the analysis to be consistent with the fundamental laws that govern relationships among the different variables. Details on these methods are described in Chapter 2.

The atmospheric reanalyses assessed in the Product provide values for all atmospheric variables over the entire globe, extending in height from the Earth's surface up to elevations of approximately 30 miles. These values provide a continuous, detailed record of how the atmosphere has evolved every 6 to 12 hours over periods spanning multiple decades. Henceforth, in this Product the term *reanalysis* refers to this specific method for reconstructing past weather and climate conditions, unless stated otherwise.

Chapter 2 describes reanalysis methods and assesses the strengths and limitations of current reanalysis products, including representations of seasonal-to-decadal climate variations and regional trends in surface temperatures and precipitation. Specific questions addressed in that Chapter are:

- What is a climate reanalysis? What role does reanalysis play within a comprehensive climate observing system?
- What can reanalysis tell us about climate processes and their representation in models used for climate predictions and climate change projections?
- What is the capacity of current reanalyses to help us identify and understand major seasonal-to-decadal climate variations, including changes in the frequency and intensity of climate extremes such as droughts?
- To what extent is there agreement or disagreement between climate trends in surface temperature and precipitation

derived from reanalyses and those derived from independent data; that is, from data that are not included in constructing the reanalysis?

- What steps would be most useful in reducing false jumps and trends in climate time series (those that may be due to changes in observing systems or other non-physical causes) and other uncertainties in past climate conditions through improved reanalysis methods? Specifically, what contributions could be made through advances in data recovery or quality control, modeling, and/or data assimilation techniques?

The assessment of capabilities and limitations of current reanalysis datasets for various purposes will be of value for determining best uses of current reanalysis products for scientific and practical purposes. This Chapter will also be useful for science program managers in developing priorities for improving the scientific and practical value of future climate reanalyses.

1.2 ATTRIBUTION

The term *attribute* has as a common use definition "to assign to a cause or source" (Webster's II Dictionary, 1988). The Intergovernmental Panel on Climate Change (IPCC) has specifically stated that "attribution of causes of climate change is the process of establishing the most likely causes for the detected change with some level of confidence" (IPCC, 2007a). The term attribution in this Product is used in the same context as the IPCC definition. However, here the scope is broadened to include observed climate variations as well as detected climate change. There are three primary reasons for expanding the scope to include climate variations: (1) climate variations often have large economic impacts on regions and communities in the United States, sometimes in the billions of dollars (NCDC, 2007); (2) there is strong public interest in explanations of the causes of major short-term climate variations, for example,

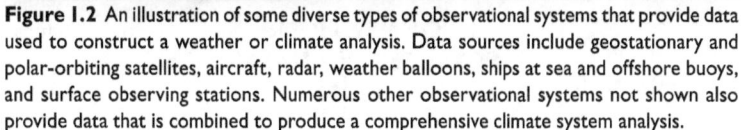
Figure 1.2 An illustration of some diverse types of observational systems that provide data used to construct a weather or climate analysis. Data sources include geostationary and polar-orbiting satellites, aircraft, radar, weather balloons, ships at sea and offshore buoys, and surface observing stations. Numerous other observational systems not shown also provide data that is combined to produce a comprehensive climate system analysis.

related to the El Niño-Southern Oscillation (ENSO), severe droughts, and other extreme events; and (3) many impacts of climate change are likely to be experienced through changes in extreme weather and climate events; that is, through changes in variability as well as changes in average conditions (IPCC, 2007b).

Methods for attributing the causes of observed climate variations and trends are discussed in Chapter 3, including the use of reanalysis data for this purpose. This Chapter focuses on observed climate variations and changes over the North American region, extending from approximately 1950 to the present, the maximum time extent of current reanalysis records. The key questions are:

- What is climate attribution? What are the scientific methods used for establishing attribution?
- What is the present understanding of the causes for North American climate trends in annual temperature and precipitation during the reanalysis record?
- What is the present understanding of causes for seasonal and regional variations in U.S. temperature and precipitation trends over the reanalysis record?

The assessment of capabilities and limitations of current reanalysis datasets for various purposes will be of value for determining best uses of current reanalysis products for scientific and practical purposes.

- What are the nature and causes of apparent rapid climate shifts that are relevant to North America over the reanalysis record?
- What is the present understanding of the causes for high-impact drought events over North America during the reanalysis record?

This Chapter will aid policy makers in assessing present scientific understanding and remaining uncertainties regarding the causes of major climate variations and trends over North America since the mid-twentieth century. Resource managers and other decision makers, as well as the general public, will also benefit from this assessment, especially for those events that have high societal, economic, or environmental impacts, such as major droughts.

1.3 CONNECTIONS BETWEEN REANALYSIS AND ATTRIBUTION

This Product focuses on two major topics: climate reanalysis and attribution. Are there scientific connections between reanalysis and attribution and, specifically, why might reanalysis be useful for determining attribution? Figure 1.3 illustrates schematically some key steps commonly used in climate science including reanalysis and attribution.

1.3.1 Steps in climate science
Observations provide the foundation for all of climate science. The observations are obtained from numerous disparate observing systems (see Figure 1.2) and are also distributed irregularly both in time and space over the Earth. These issues and others pose significant challenges to scientists in evaluating present climate conditions and in comparing present conditions with those of the past.

As discussed previously, an analysis is a method for combining diverse observations to obtain a quantitative (numerical) depiction of the state of the atmosphere or, more generally, the state of the climate system at a given time (Figure 1.3). Reanalysis corresponds to the step of applying the same analysis method to carefully reconstruct the past climate history. Extending the record back in time enables scientists to detect climate variations and changes, and to compare present and past conditions. This reanalysis must apply consistent methods and quality-controlled data in order to accurately identify changes over time and determine how changes in different variables such as winds, temperatures, and precipitation are related. In climate science, attribution corresponds to what in medical science is called diagnosis; that is, it is the process of identifying the cause or causes of the feature of interest. As in medical science, additional "diagnostic tests" are often required to establish attribution. In climate science, these additional tests most commonly consist of controlled experiments conducted with climate models; results are compared between model outcomes when a climate forcing of interest (*e.g.*, from changes in greenhouse gases or volcanic aerosols) is either included or excluded in order to assess its potential effects.

Establishing attribution provides a scientific underpinning for predicting future climate and information useful for evaluating needs and options for adaptation and/or mitigation due to climate variability or change. Detailed discussions of climate prediction, adaptation, and mitigation are beyond the scope of this

Reanalysis can be considered as playing a central role in determining *what* has happened in the climate system, while attribution is necessary to address the question of *why* the changes have occurred.

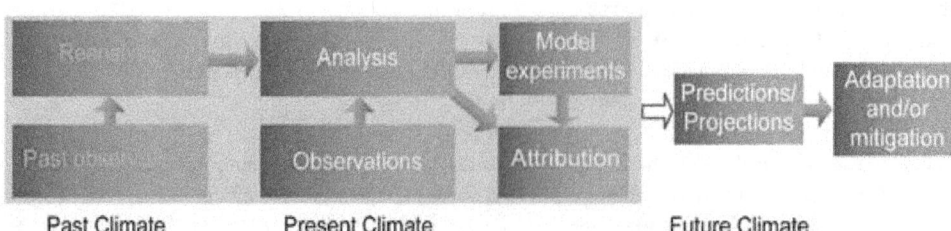

Past Climate   Present Climate   Future Climate

Figure 1.3 Schematic illustrating some key steps in climate science. The shaded box indicates the general scope of this Product. The arrows show the general flow of information leading to decision support for adaptation and/or mitigation. As indicated by the open arrow, the description and understanding of past and current climate conditions included within the shaded box provide key information for developing reliable predictions of future climate and for evaluating options for adaptation and/or mitigation.

Product; however, recognition of such relationships helps illuminate potential applications of, and connections between, climate reanalysis and attribution.

1.3.2 Further comments

Reanalysis can be considered as playing a central role in determining *what* has happened in the climate system (what has changed, and by how much?), while attribution is necessary to address the question of *why* the changes have occurred. As illustrated by Figure 1.3, observations serve as the fundamental starting point for climate reanalysis; observations themselves are generally not sufficient to establish attribution; models incorporating fundamental understanding of key physical processes and their relationships are also required. The event of interest, *e.g.*, a long-term trend or other feature, such as a severe drought, must first be identified with confidence in the data in order for attribution to be meaningful. Reanalysis often plays an important role in this regard by providing a comprehensive, high quality, and continuous climate dataset spanning several decades. Physical consistency, obtained through the use of a model that incorporates the fundamental governing laws of the climate system, is also a primary feature of reanalysis datasets. Physical consistency enables identification of the roles of various processes in producing climate variations and change along with corresponding linked patterns of variability. Thus, the method of reanalysis can contribute to more confident attribution of the processes that produce responses within the climate system to a given climate forcing, as well as the expected geographical patterns and magnitudes of the responses.

One potential application of reanalysis data is in the detection of climate change. Within the IPCC, detection of climate change is defined as the process of demonstrating that climate has changed in some defined statistical sense, *without providing a reason for that change.* As stated earlier, attribution of the causes of climate change is the process of establishing the most likely cause for the detected change with some level of confidence. Reanalysis can play an important role in both detecting and attributing causes of climate variations and change; however, it is vital to recognize that

reanalysis alone is seldom sufficient and that the best methods for both detection and attribution often depend on results obtained from a broad range of datasets, models, and analysis techniques.

In order to establish more definitive attribution, climate scientists perform controlled climate model experiments to determine whether estimated responses to particular climate forcings are consistent with the observed climate features of interest (*e.g.*, a sustained temperature trend or a drought). Reanalysis data can also be of considerable value in evaluating how well climate models represent observed climate features and responses to different forcings over several decades, thereby providing important guidance of the utility of the models for establishing attribution.

There are inevitable uncertainties associated with observational data, analysis techniques, and climate models. Therefore, climate change detection and attribution findings must be stated in probabilistic terms based on current knowledge, and expert judgment is often required to assess the evidence regarding particular processes (see Chapter 3). The language on uncertainty adopted in this Product is consistent with the IPCC Fourth Assessment Report (IPCC, 2007a). Finally, it is important to recognize that in complex systems, whether physical, biological, or human, it is often not one factor but the interaction among multiple factors that determines the ultimate outcome.

1.4 REANALYSIS APPLICATIONS AND USES

Over the past several years, reanalysis datasets have become a cornerstone for research in advancing our understanding of how and why climate has varied since the mid-twentieth century. For example, Kalnay *et al.* (1996), the initial overview paper on one of the first reanalysis datasets produced in the United States, has been cited more than 5500 times in the peer-reviewed literature as of mid-2008 and is currently the most widely cited paper in the geophysical sciences (ISI Web of Knowledge, <http://www.isiwebofknowledge.com/>; see also Figure 1.4).

Over the past several years, reanalysis datasets have become a cornerstone for research in advancing our understanding of how and why climate has varied since the mid-twentieth century.

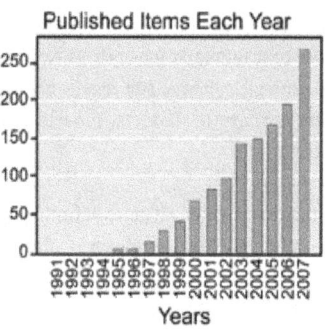
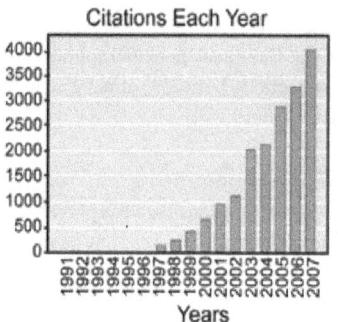

Figure I.4 The number of published items and citations from an "ISI Web of Science" search with the key words REANALYSIS and CLIMATE.

Reanalysis datasets are increasingly used for practical applications in sectors such as energy, agriculture, water resources, and insurance.

Increasingly, reanalysis data are being used in a wide range of practical applications. One important application is to address the question: "How is the present climate similar to, or different from, past conditions?" The short time intervals of reanalysis data (typically, every 6 to 12 hours) enable detailed studies of the time evolution of specific weather and climate events as well as comparisons with similar events in the past, providing important clues on key physical processes. Intercomparisons of different reanalyses and observational datasets help to provide a measure of the uncertainty in representations of past climate, including identifying phenomena, regions, and time periods for which confidence in features is relatively high or low (Santer *et al.*, 2005).

Reanalysis datasets are also increasingly used for practical applications in sectors such as energy, agriculture, water resources, and insurance (*e.g.*, Schwartz and George, 1998; Pryor *et al.*, 2006; Challinor *et al.*, 2005; Pulwarty, 2003; Pinto *et al.*, 2007). For example, a recently completed high-resolution regional reanalysis, the North American Regional Reanalysis (Mesinger *et al.*, 2006), focuses on improving the representation of the water cycle over North America in order to better serve water resource management needs. Chapter 2 of this Product will inform users of strengths and limitations of current reanalysis datasets, and aid in determining whether certain datasets are suited for specific purposes. Chapter 3 will be of value to policy-makers and the public in providing an assessment of current scientific understanding on the causes for observed climate variations and change over North America from the mid-twentieth century to the present. Finally, Chapter 4 recommends steps needed to

improve national capabilities in reanalysis and attribution in order to increase their value for scientific and practical applications.

CHAPTER 2

Reanalysis of Historical Climate Data for Key Atmospheric Features

Convening Lead Author: Siegfried Schubert, NASA

Lead Authors: Phil Arkin, Univ. of Maryland; James Carton, Univ. of Maryland; Eugenia Kalnay, Univ. of Maryland; Randal Koster, NASA

Contributing Authors: Randall Dole, NOAA; Roger Pulwarty, NOAA

KEY FINDINGS

- Reanalysis plays a crucial integrating role within a global climate observing system by producing comprehensive, long-term, objective, and consistent records of climate system components, including the atmosphere, oceans, and land surface (Section 2.1).

- Reanalysis data play a fundamental and unique role in studies that address the nature, causes, and impacts of global-scale and regional-scale climate phenomena (Section 2.3).

- Reanalysis datasets are of particular value in studies of the physical processes that produce high-impact weather and climate events such as droughts and floods, as well as other key atmospheric features that affect the United States, including climate variations associated with major modes of climate variability, such as the El Niño-Southern Oscillation (Section 2.3).

- Global and regional surface temperature trends in reanalysis datasets are broadly consistent with those obtained from temperature datasets constructed from surface observations not included in the reanalysis, particularly since the late 1970s. However, in some regions (e.g., Australia) the reanalysis trends show major differences with observations (Section 2.4).

- Reanalysis precipitation trends are less consistent with those calculated from observational datasets. The differences are likely due principally to current limitations in the reanalysis models and the methods used for integrating diverse datasets within models (Section 2.4).

- Current reanalysis data are extremely valuable for a host of scientific and practical applications; however, the overall quality of reanalysis products varies with latitude, altitude, time period, location and time scale, and quantity or variable of interest (Sections 2.1, 2.3).

- Current global reanalysis data are most reliable in Northern Hemisphere mid-latitudes, in the middle to upper troposphere (about 3 to 12 miles above Earth's surface), and for regional and larger areas. They are also most reliable for time periods ranging from one day up to several years, making reanalysis data well suited for studies of mid-latitude storms and short-term climate variability (Sections 2.1, 2.2, 2.3, 2.4).

- Present reanalyses are more limited in their value for detecting long-term climate trends, although there are cases where reanalyses have been usefully applied for this purpose. Important factors constraining the value of reanalyses for trend detection include: changes in observing systems over time; deficiencies in observational data quality and spatial coverage; model limitations in representing interactions across the land-atmosphere and ocean-atmosphere interfaces, which affect the quality of surface and near-surface weather and climate variables; and inadequate representation of the water cycle (Sections 2.2, 2.3, 2.4).

- At the present time, data sets constructed for an individual variable, for example, surface temperature or precipitation, are generally superior for climate change detection. However, the integrated and comprehensive nature of reanalysis data provides a quantitative foundation for improving understanding of the processes that produce changes. These qualities make reanalysis data more useful than individual variable datasets for attributing the causes of climate variations and change (Section 2.4).

- Reanalysis data play an important role in assessing the ability of climate models to simulate basic weather and climate variables such as the horizontal winds, temperature, and pressure. In addition, the adjustments or analysis increments produced during the course of a reanalysis provide a method to identify fundamental errors in the physical processes and/or missing physics that create climate model biases (Sections 2.2, 2.3).

- Reanalyses have had substantial benefits for climate research and prediction, as well as for a wide range of societal applications. Significant future improvements are possible by developing new methods to address observing system inconsistencies, by developing estimates of the reanalysis uncertainties, by improving the observational database, and by developing integrated Earth system models and analysis systems that incorporate key climate elements not included in atmospheric reanalyses to date (Section 2.5).

2.1. CLIMATE REANALYSIS AND ITS ROLE WITHIN A COMPREHENSIVE CLIMATE OBSERVING SYSTEM

2.1.1 Introduction

Weather and climate vary continuously around the world on all time scales. The observation and prediction of these variations is important to many aspects of human society. Extreme weather events can cause significant loss of life and damage to property. Seasonal to interannual changes associated with the El Niño-Southern Oscillation (ENSO) phenomenon and other modes of climate variability have substantial effects on the economy. Climate change, whether natural or anthropogenic, can profoundly influence social and natural environments throughout the world, with impacts that can be large and far-reaching.

Determining the nature and predictability of climate variability and change is crucial to society's future welfare.

Determining the nature and predictability of climate variability and change is crucial to society's future welfare. To address the threats and opportunities associated with weather phenomena, an extensive weather observing system has been put in place over the past century (see Figure 2.1). Considerable resources have been invested in obtaining observations of the ocean, land, and atmosphere from satellite and surface-based systems, with plans to improve and expand these observations as a part of the Global Earth Observing System of Systems (GEOSS, 2005). Within this developing climate observing system, climate analysis plays an essential synthesizing role by combining data obtained from this diverse array of Earth system observations to enable improved descriptions and understanding of climate variations and change.

2.1.2 What is a Climate Analysis?

As discussed in Chapter 1, at its most fundamental level, an *analysis* is a detailed representation of the state of the atmosphere and, more generally, of other Earth climate system components, such as oceans or land surface, that is based on observations. A number of techniques can be used to create an analysis from a given set of observations.

One common technique for creating an analysis is based on the expertise of human analysts, who apply their knowledge of phenomena and physical relationships to estimate values of variables between observation locations, a technique referred to as interpolation. Such subjective analysis methods were used almost exclusively before the onset of modern numerical weather prediction in the 1950s and are still used for many purposes today. While these techniques have certain advantages, including the relative simplicity by which they may be produced, there are key inadequacies that limit their value for numerical weather prediction and climate research. An important practical limitation, recognized in the earliest attempts at numerical weather prediction (Richardson, 1922; Charney, 1951), was that the process of creating a detailed analysis, for example, of the global winds and temperatures through the depth of the atmosphere on a given day, is

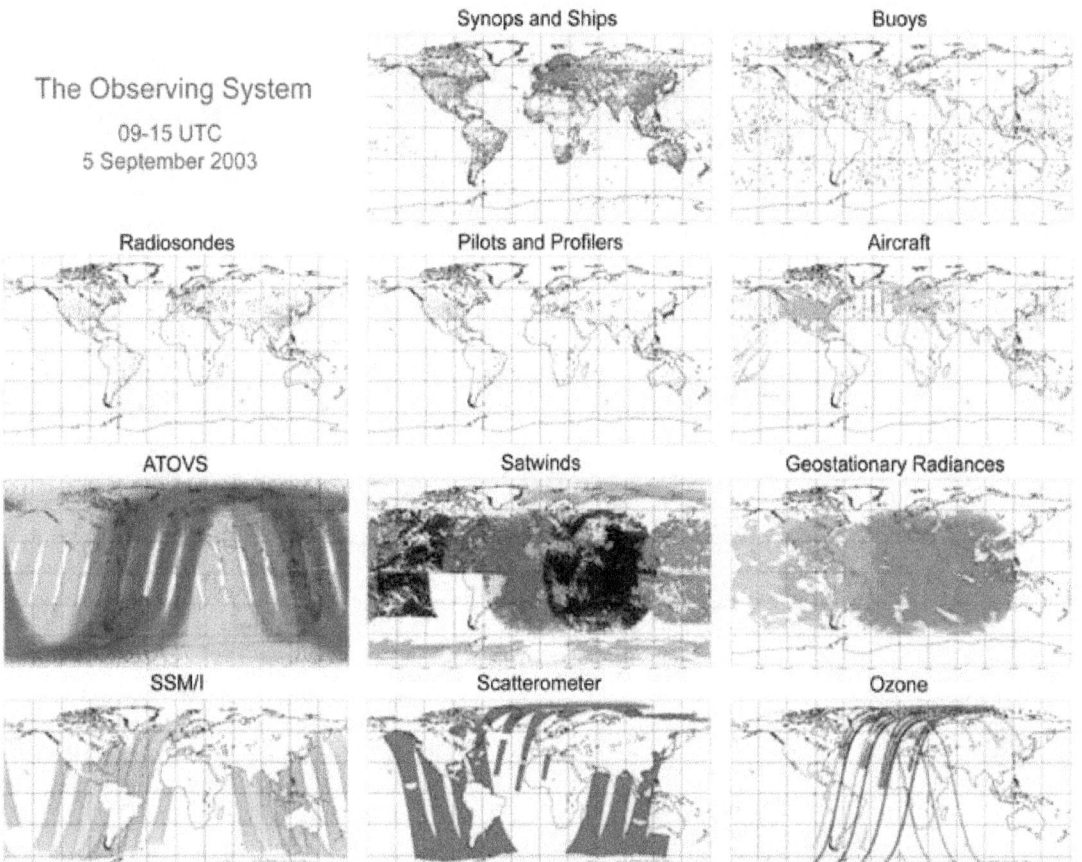

Figure 2.1 The atmospheric data coverage provided by the modern observing systems on 5 September 2003 for use in reanalysis. From Simmons (2006).

time consuming, often taking much longer to produce than the evolution of the weather itself. A second limitation is that physical imbalances between fields that are inevitably produced during a subjective analysis lead to forecast changes that are much larger than actually observed (Richardson, 1922). A third limitation is that this type of subjective analysis is not reproducible. In other words, the same analyst, given the same observational data, will generally not produce an identical analysis when given multiple opportunities.

Thus, by the early 1950s the need for an automatic, objective analysis of atmospheric conditions had become apparent. The important technological advance provided by the early computers of that time, while primitive by today's standards, could still perform calculations far faster than previously possible, making this a feasible goal.

The first objective analyses used simple statistical techniques to interpolate data values from the locations where observations were made onto uniform spatial grids that were used for the model predictions. Such techniques are still widely employed today to produce many types of analyses, such as global maps of surface temperatures, sea surface temperature (SST), and precipitation (Jones *et al.*, 1999; Hansen *et al.*, 2001; Doherty *et al.*, 1999; Huffman *et al.*, 1997; Xie and Arkin, 1997; Adler *et al.*, 2003; Fan and Van den Dool, 2008). The purely statistical approaches are less well suited for the analysis of upper air conditions in that they do not fully exploit known physical relationships among different variables of the climate system, for example, among fields of temperature, winds, and atmospheric pressure. These relationships place fundamental constraints on how weather and climate evolve in time. Therefore, statistical analysis techniques are no longer used for applications that depend on relationships among

The first objective analyses used simple statistical techniques to interpolate data values from the locations where observations were made onto uniform spatial grids that were used for the model predictions.

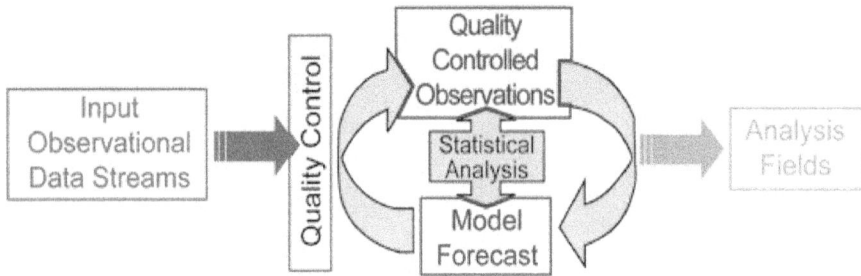

Figure 2.2 A schematic of data assimilation.

variables, as in numerical weather prediction or in research to assess detailed mechanisms for climate variability and change.

An alternative objective analysis method, which is the principal focus for this Product, is to estimate the state of the climate system (or of one of its components) by combining observations together within a numerical prediction model that mathematically represents the physi-

cal and dynamical processes operating within the system. This observations-model integration is achieved through a technique called data assimilation. One important aspect of a comprehensive climate observing system achieved through data assimilation is the ability to integrate diverse surface, upper air, satellite, and other observations together into a coherent, consistent description of the state of the global climate system. Figure 2.1 shows, for example, a snapshot of the coverage provided by the different atmospheric observing systems on 5 September 2003 that can be incorporated into such an analysis scheme.

How are observations combined that have such different spatial coverage, sampling density, and error characteristics? Data assimilation mathematically combines a background field or an initial estimate produced by a numerical prediction of the atmosphere (or oceans) with available observations using a method designed to minimize the overall errors in the analysis. Figure 2.2 schematically shows how data assimilation combines quality-controlled observations with a short-term model forecast (typically, in six-hour increments) to produce an analysis that attempts to minimize errors in estimates of the atmospheric state that would be present due to either the observations or model evaluated separately (for more details see Appendix A).

In practice, the quality of a global analysis is impacted by a multitude of practical decisions and compromises, involving the analysis methodology, quality control, the choice of observations and how they are used, and the model (see Appendix A and the discussion below). Figure 2.3 compares three different reanalyses produced from the observations available for 5 September 2003 (Figure 2.1) of the 500 millibars (mb) geopotential height distribution (the height of a mid-tropospheric pressure surface above mean sea level) and total water vapor fields. These are results from the National Centers for Environmental Prediction (NCEP)/National Center

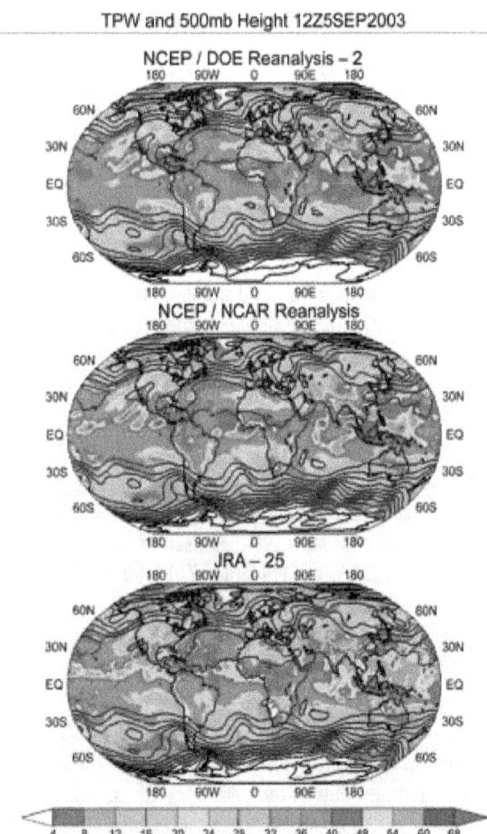

Figure 2.3 The global distribution of the mid-tropospheric pressure field (contours are of the 500 millibars [mb] geopotential height field) and total water vapor (shaded color; units are in millimeters) for 5 September 2003 from three different analyses.

Table 2.1 Characteristics of existing atmospheric reanlyses.

Organization	Time Period	Model	Analysis Scheme	Output	References
NASA Data Asssimilation Office (DAO)	1980 to 1994	2X2.5° Lat/lon-Δx~250 km, L20 (σ, top at 10mb), specified soil moisture	Optimal Interpolation (OI) with incremental analysis update	No longer available	Schubert *et al.* (1993)
NOAA NCEP and NCAR (R1)	1948 to present	T62 - Δx~200km L28 (σ, top at about 3mb)	Spectral Statistical Interpolation (SSI)	<http://www.cpc.ncep.noaa.gov/products/wesley/reanalysis.html>	Kalnay *et al.* (1996)
NOAA NCEP and DOE (R2)	1979 to present	T62 - Δx~200km L28 (σ, top at about 3mb)	Spectral Statistical Interpolation (SSI)	<http://www.cpc.ncep.noaa.gov/products/wesley/reanalysis2/>	Kanamitsu *et al.* (2002) (Fixes errors found in R1 including fixes to PAOBS, snow, humidity, etc.)
European Centre for Medium-Range Weather Forecasts (ECMWF) Reanalysis (ERA-15)	1979 to 1993	T106 - Δx~125km L31(σ-p, top at 10mb)	Optimal Interpolation (OI), IDVAR, nonlinear normal mode initialization	<http://data.ecmwf.int/data/d/era15/>	Gibson *et al.* (1997)
ECMWF (ERA-40)	1957 to 2001	T159 - Δx~100km L60 (σ-p, top at 0.1mb)	3D-Var, radiance assimilation	<http://data.ecmwf.int/data/d/era40_daily/>	Uppala *et al.* (2005)
JMA and CRIEPI (JRA-25)	1979 to 2004	T106 - Δx~125km L40 (σ-p, top at 0.4mb)	3D-Var, radiance assimilation	<http://jra.kishou.go.jp/index_en.html>	Onogi *et al.* (2005)
NOAA North American Regional Reanalysis (NARR)	1979 to present	Δx= 32km L45	3D-Var, precipitation assimilation	<http://nomads.ncdc.noaa.gov/#narr_data sets>	Mesinger *et al.* (2006)

for Atmospheric Research (NCAR) Reanalysis 1, the NCEP/Department of Energy (DOE) Reanalysis 2, and the Japanese Meteorological Agency (JMA)/Central Research Institute of Electrical Power Industry (CRIEPI) 25-year Japanese Reanalysis (JRA-25).

The two NCEP reanalyses were carried out with basically the same system (Table 2.1, the NCEP/DOE reanalysis system corrected some of the known errors in the NCEP/NCAR system).

The three analyses show substantial agreement in midlatitudes, especially for the pressure distribution; however, there is substantial disagreement in the tropical moisture fields between the NCEP and JRA data. The differences indicate that there are insufficient observations and/or inadequate representation of relevant physical processes incorporated into the models that are needed to tightly constrain the analyses.

Consequently, the uncertainties in the tropical moisture field are relatively large.

The numerical prediction model used for data assimilation plays a fundamental role in the analysis. It ensures an internal consistency of physical relationships among variables such as temperatures, pressure, and wind fields, and provides a detailed, three-dimensional representation of the system state at any given time, including winds, temperatures, pressures, humidity, and numerous other variables that are necessary for describing weather and climate (Appendix A). Further, the physical relationships among atmospheric (or oceanic) variables that are represented in the mathematical model enable the model to transfer information from times or regions with more observations to other times or areas with sparse observations. At the same time, potential errors are introduced by the use of a model (Section 2.2).

The numerical prediction model used for data assimilation plays a fundamental role in the analysis.

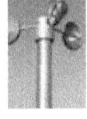

Beginning in the 1970s, the sequence of initial atmospheric conditions or analyses needed for the emerging comprehensive global numerical weather prediction models were also used to study climate (Blackmon *et al.*, 1977; Lau *et al.*, 1978; Arkin, 1982). This unforeseen use of the analyses marked what could be considered a revolutionary step forward in climate science, enabling for the first time detailed quantitative analyses that were instrumental in advancing the identification, description, and understanding of many large scale climate variations, in particular, some of the major modes of climate variability described in Section 2.3. However, the frequent changes in analysis systems (*e.g.,* model upgrades) needed to improve short-range numerical weather forecasts also introduced false shifts in the perceived climate that rendered these initial analyses unsuitable for problems such as detecting subtle climate trends. Recognition of this fundamental issue led to recommendations for the development of a comprehensive, consistent analysis of the climate system, effectively introducing the concept of a model-based climate reanalysis (Bengtsson and Shukla, 1988; Trenberth and Olson, 1988).

2.1.3 What is a Climate Reanalysis?

A climate reanalysis is an analysis performed with a fixed (*i.e.*, not changing in time) numerical prediction model and data assimilation method that assimilates quality-controlled observational data over an extended time period, typically several decades, to create a long-period climate record. This use of a fixed model and data assimilation scheme differs from analyses performed for daily weather prediction. Such analyses are conducted with models using numerical and/or physical formulations as well as data assimilation schemes that are updated frequently, sometimes several times a year, giving rise to false changes in climate that limit their value for climate applications. Climate analysis also fundamentally differs from weather analysis in that observations throughout the system evolution are available for use, rather than simply those observations made immediately prior to the time when the forecast

Current methods of climate reanalyses evolved from methods developed for short-range weather prediction, and have yet to realize their full potential for climate applications.

is initiated. While weather analysis has the goal of enabling the best short-term weather forecasts, climate analysis can be optimized to achieve other objectives such as providing a consistent description of the atmosphere over an extended time period. Current methods of climate reanalyses evolved from methods developed for short-range weather prediction, and have yet to realize their full potential for climate applications (see Chapter 4).

In the late 1980s, several reanalysis projects were initiated to develop long-term records of analyses better suited for climate purposes (Table 2.1). The products of these first reanalyses (*e.g.*, maps of daily, monthly, and seasonal averages of temperatures, winds, and humidity) have proven to be among the most valuable and widely used in the history of climate science, as indicated both by the number of scholarly publications that rely upon them and by their widespread use in current climate services (see Section 1.4). The reanalysis projects have produced detailed atmospheric climate records that have enabled successful climate monitoring and research to be conducted. They have also provided a testbed for improving prediction models on all time scales (see Section 2.2), especially for seasonal-to-interannual forecasts, as well as greatly improved basic observations and datasets prepared for their production. When extended to the present as an ongoing climate analysis, reanalysis provides decision makers with information about current climate events in relation to past events, and contributes directly to climate change assessments.

2.1.4 What Role Does Reanalysis Play within a Climate Observing System?

One of the key limitations of current and foreseeable observing systems is that they do not provide complete spatial coverage of all relevant components of the climate system. Because the observing system has evolved over the last half century mainly in response to numerical weather prediction needs, it is focused primarily on the atmosphere. The system today consists of a mixture of *in situ* and remotely sensed observations with differing spatial and temporal sampling and error characteristics (Figure 2.1). An example of the observations available for reanalysis during the modern satellite era is provided in Table 2.2.

A major strength of modern data assimilation methods is the use of a model to help fill in the gaps of the observing system. The assimilation methods act as sophisticated interpolators that use the complex equations governing the atmosphere's evolution together with all available observations to estimate the state of the atmosphere in regions with little or no observational coverage. Statistical schemes are used that ensure that, in the absence of bias with respect to the true state of the atmosphere, the observations and model first guess are combined in an optimal way to jointly minimize errors that are subject to certain simplifying assumptions such that the statistics follow a normal distribution. This can be as simple as the model transporting

> A major strength of modern data assimilation methods is the use of a model to help fill in the gaps of the observing system.

Table 2.2 An example of the conventional and satellite radiance data available for reanalysis during the satellite era (late 1970s to present). These are the observations used in the new NASA Modern Era Retrospective-Analysis for Research and Applications (MERRA) reanalysis (Section 2.5.2).

Data Source/Type	Period	Data Supplier
Conventional Data		
Radiosondes	1970 to present	NOAA/NCEP
PIBAL winds	1970 to present	NOAA/NCEP
Wind profiles	1992/5/14 to present	UCAR CDAS
Convetional, ASDAR, and MDCRS aircraft reports	1970 to present	NOAA/NCEP
Dropsondes	1970 to present	NOAA/NCEP
PAOB	1978 to present	NCEP CDAS
GMS, METEOSAT, cloud drift IR and visible winds	1977 to present	NOAA/NCEP
GOES cloud drift winds	1997 to present	NOAA/NCEP
EOS/Terra/MODIS winds	2002/7/01 to present	NOAA/NCEP
EOS/Aqua/MODIS winds	2003/9/01 to present	NOAA/NCEP
Surface land observations	1970 to present	NOAA/NCEP
Surface ship and buoy observations	1977 to present	NOAA/NCEP
SSM/I rain rate	1987/7 to present	NASA/GSFC
SSM/I V6 wind speed	1987/7 to present	RSS
TMI rain rate	1997/12 to present	NASA/GSFC
QuikSCAT surface winds	1999/7 to present	JPL
ERS-1 surface winds	1991/8/5 to 1996/5/21	CERSAT
ERS-2 surface winds	1996/3/19 to 2001/1/17	CERSAT
Satellite Data		
TOVS (TIROS N, N-6, N-7, N-8)	1978/10/30 to 1985/01/01	NCAR
(A)TOVS (N-9, N-10, N-11, N-12)	1985/01/01 to 1997/07/14	NOAA/NESDIS & NCAR
(A)TOVS (N-14, N-15, N-16, N-17, N-18)	1995/01/19 to present	NOAA/NESDIS
EOS/Aqua	2002/10 to present	NOAA/NESDIS
SSM/I V6 (F08, F10, F11, F13, F14, F15)	1987/7 to present	RSS
GOES sounder T_B	2001/01 to present	NOAA/NCEP
SBUV2 ozone (Version 8 retrievals)	1978/10 to present	NASA/GSFC/Code 613.3

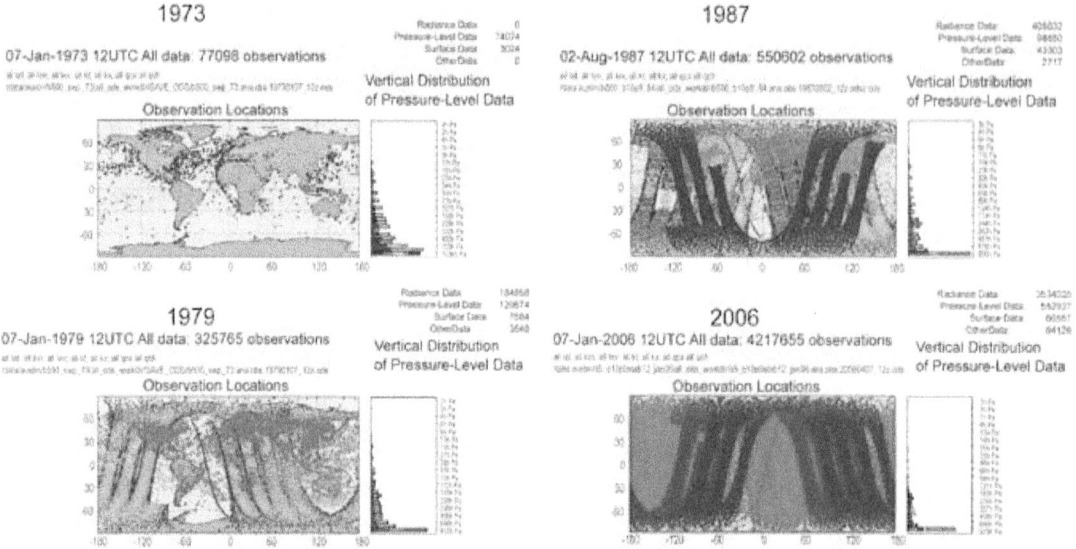

Figure 2.4 Changes in the distribution and number of observations available for NASA's Modern Era Retrospective-Analysis for Research and Applications (MERRA) reanalysis.

The use of a model enables estimates of quantities and physical processes that are difficult to observe directly, such as vertical motions, surface heat fluxes, latent heating, and many of the other physical processes that determine how the atmosphere evolves over time.

warm air from a region that has good observational coverage (*e.g.*, over the United States) to a region that has little or no coverage (*e.g.*, over the adjacent ocean), or a more complicated example, where the model generates a realistic low-level jet in a region where such phenomena exist but observations are limited. The latter is an example of a phenomenon that is largely generated by the model, and only indirectly constrained by observations. This example highlights both the advantages and difficulties in using reanalysis for climate studies. Through the use of a model, it allows climate scientists to estimate features that are indirectly or incompletely measured; however, the scientists have confidence in those estimates only if they are able to account for all model errors.

The use of a model also enables estimates of quantities and physical processes that are difficult to observe directly, such as vertical motions, surface heat fluxes, latent heating, and many of the other physical processes that determine how the atmosphere evolves over time. In general, the estimated quantities are model dependent and careful interpretation is required. Any incorrect representation of physical processes (called parameterizations) will be reflected in the reanalysis to some extent. Only recently have the models improved enough to be used with some confidence in individual physical processes. Previously, most studies

using assimilated data have indirectly estimated physical processes by computing them as a residual of a budget that involves only variables that are well observed (Section 3.2.3). Thus, it is important to understand which quantities are strongly constrained by the observations, and which are indirectly constrained and depend on model parameterizations. In recognition of this problem, efforts have been made to document the quality of the individual products and categorize them according to how strongly they are observationally constrained (*e.g.*, Kalnay *et al.*, 1996; Kistler *et al.*, 2001).

Beyond their fundamental integrating role within a comprehensive climate observing system, climate analysis and reanalysis can also be used to identify redundancies and gaps in the climate observing system, thus enabling the entire system to be configured more cost effectively. By directly linking products to observations, a reanalysis can be applied in conjunction with other science methods to optimize the design and efficiency of future climate observing systems and to improve the products that the system produces.

Current reanalysis data are extremely valuable for a host of climate applications. However, there are also limitations. These are due, for example, to changes in the observing systems, such as the substantial increase in satellite

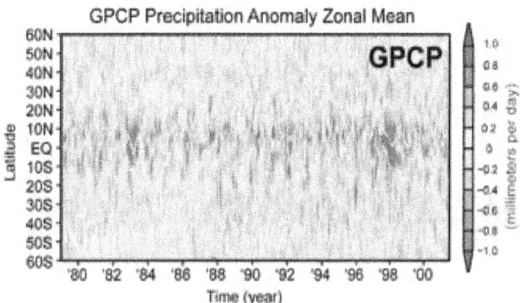

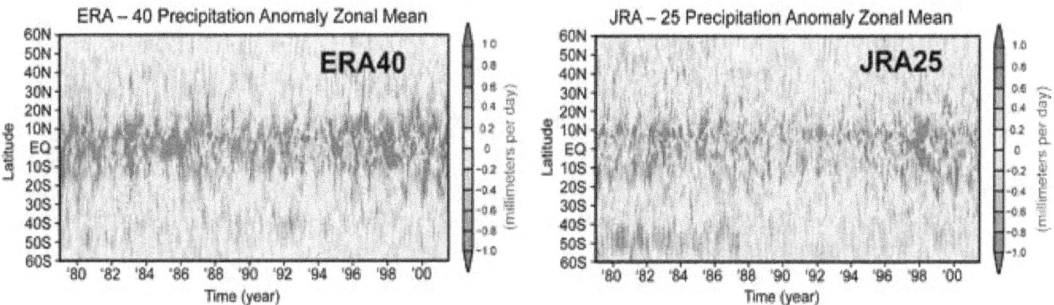

Figure 2.5 Trends and shifts in the reanalyses. The figures show the zonal mean precipitation from the GPCP observations (top panel), the ERA-40 reanalysis (bottom left panel), and the JRA-25 reanalysis (bottom right panel). Courtesy of Junye Chen and Michael Bosilovich, NASA Global Modeling and Assimilation Office (GMAO).

data in 1979 and other newer remote sensing instruments (Figure 2.4). Such changes to the observing system influence the variability that is inferred from reanalyses. Therefore, inferred trends and low frequency (*e.g.*, decadal) variability may be less reliable than shorter-term weather and climate variations (*e.g.*, Figure 2.5 and discussion in Sections 2.3.2.2 and 2.4.2).

The need to periodically update the climate record in order to provide improved reanalyses for climate research and applications has been strongly emphasized (*e.g.*, Trenberth *et al.*, 2002b; Bengtsson *et al.*, 2004a). There are several reasons for these updates: (1) to include important or extensive additional observations missed in earlier analyses; (2) to correct observational data errors identified through subsequent quality-control efforts; and (3) to take advantage of scientific advances in models and data assimilation techniques, including bias correction techniques (Dee, 2005), and to incorporate new types of observations, such as satellite data not assimilated in earlier analyses. In the following Sections, the strengths and limitations of current reanalyses for address-

ing specific questions defined in the Preface are discussed.

2.2 ROLE OF REANALYSIS IN UNDERSTANDING CLIMATE PROCESSES AND EVALUATING CLIMATE MODELS

2.2.1 Introduction
Global atmospheric data assimilation combines various observations of the atmosphere (see Figure 2.1) with a short-term model forecast to produce an improved estimate of the state of the atmosphere. The model used in the assimilation incorporates current scientific understanding of how the atmosphere (and more generally the climate system) behaves and can ideally forecast or simulate all aspects of the atmosphere at all locations around the world.

Atmospheric data assimilation and reanalysis, in particular, can be thought of as a model simulation of past atmospheric behavior that is continually updated or adjusted by available observations. Such adjustments are necessary because the model would otherwise evolve differently from nature since it is imperfect (*i.e.*,

Atmospheric data assimilation and reanalysis, in particular, can be thought of as a model simulation of past atmospheric behavior that is continually updated or adjusted by available observations.

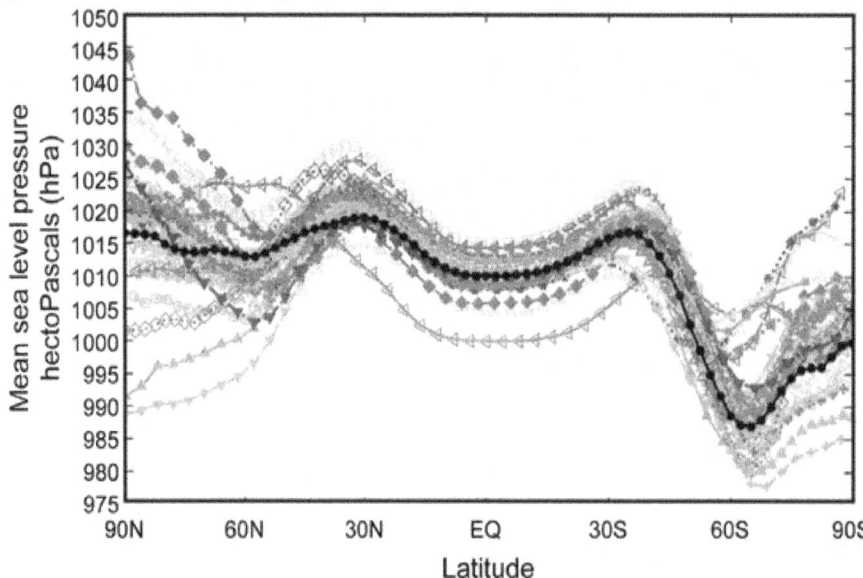

Figure 2.6 The distribution of zonally-averaged sea level pressure simulated by the various AMIP models for December, January, and February from 1979 to 1988 compared against the ECMWF (ERA-15) reanalysis (the black dots; Gibson et al., 1997). From Gates et al., 1999.

Even with a perfect model and nearly perfect observations, adjustments are necessary because the model would still deviate from nature since the atmosphere is chaotic and even very small observational errors grow rapidly to impact the model forecast.

our understanding about how the atmosphere behaves and our ability to represent that behavior in computer models is limited). The adjustments must be made continually (or at least intermittently) because the information (observations) used to correct the model's time evolution at any instant are incomplete and also contain errors. In other words, all aspects of the climate system cannot be perfectly measured. Even with a perfect model and nearly perfect observations, adjustments are necessary because the model would still deviate from nature since the atmosphere is chaotic and even very small observational errors grow rapidly to impact the model forecast.

The above model-centric view of data assimilation is useful when trying to understand how reanalysis data can be applied to evaluate how well climate models represent atmospheric processes. It highlights the fact that reanalysis products are a mixture of observations and model forecasts, and their quality will therefore be impacted by the quality of the model. In large geographic regions with little observational coverage, a reanalysis will tend to move away from nature and reflect more of the model's own behavior. Also, poorly observed quantities, such as surface evaporation, depend on the quality of the model's representation or parameteriza-

tions of the relevant physical processes (e.g., the model's land surface and cloud schemes). Given that models are an integral component of reanalysis systems, how then can reanalyses be used to help understand errors in the climate models—in some cases the same models used to produce the reanalysis?

2.2.2 Assessing Systematic Errors
The most straightforward approach to assessing systematic errors is to compare the basic reanalysis conditions (e.g., winds, temperature, moisture) with those that the model produces in free-running mode (a simulation that is not corrected by observations)[1]. The results of such comparisons, for example of monthly or seasonal average values, can indicate whether the model has systematic errors such as producing too cold or too wet in certain regions.

In general, such comparisons are only useful for regions and for quantities where the uncertainties in the reanalysis data are small compared to the model errors. For example, if the difference in the tropical moisture between two reanalysis products (e.g., NCEP/NCAR R1 and ERA-40) is as large as (or larger than) the

---

[1] These are typically multi-year Atmospheric General Circulation Model runs started from arbitrary initial conditions and forced by the observed record of sea surface temperatures (SST).

differences between any one reanalysis product and the model results, then no conclusion can be reached about the model quality based on that comparison. This points to the need for obtaining reliable uncertainty and bias estimates of all reanalysis quantities (*e.g.*, Dee and Todling, 2000), something that has not yet been achieved in the current generation of reanalysis efforts. In the absence of such estimates, comparing the available reanalysis datasets can provide guidance regarding uncertainties and model dependence. Such comparisons with reanalysis data are now routine and critical aspects of any model development and evaluation effort. (*e.g.*, Atmospheric Model Intercomparison Project [AMIP] [Gates, 1992], the tropospheric-stratospheric GCM-Reality Intercomparison Project for SPARC [GRIPS] [Pawson *et al.*, 2000], and coupled model evaluation conducted for the IPCC Fourth Assessment Report [IPCC, 2007]).

Figure 2.6 illustrates a comparison between various atmospheric models and the first European Centre for Medium-Range Weather Forecasts (ECMWF) reanalysis (ERA-15, Table 2.1).

The comparison shows considerable differences among the models in the zonal mean surface pressure, especially at high latitudes. Figure 2.7 shows an example of a more in-depth evaluation of the ability of Atmospheric General Circulation Model (AGCM) simulations forced by observed sea surface temperatures to reproduce that part of the variability associated with ENSO.

In this case the comparison is made with the NCEP/NCAR R1 reanalysis for December, January, and February

from 1950 to 1999. The comparison suggests that the models produce a very good response to the ENSO-related sea surface temperature variations.

2.2.3 Inferences about Climate Forcing

While the above comparisons address errors in the description of the climate system, a more challenging problem is to address errors in the forcing or physical mechanisms (in particular the parameterizations) by which the model produces and maintains climate anomalies. This involves quantities that are generally only weakly or indirectly constrained by observations (*e.g.*, Kalnay *et al.*, 1996; Kistler *et al.*, 2001). Ruiz-Barradas and Nigam (2005), for example, show that land/atmosphere interactions may be too efficient (make too large a contribution) in maintaining precipitation anomalies in the U.S. Great Plains in current climate models, despite rather substantial differences in the reanalyses. Nigam and Ruiz-Barradas (2006) highlight some of the difficulties encountered when trying to validate models in the presence

> Comparing available reanalysis datasets can provide guidance regarding uncertainties and model dependence.

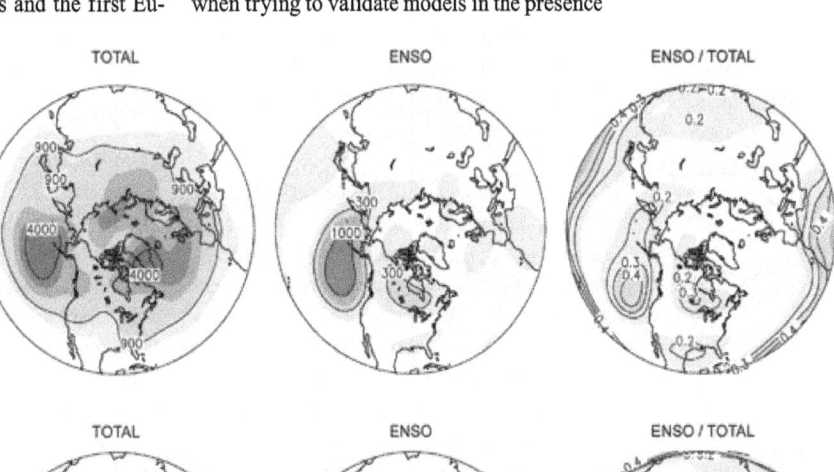

Figure 2.7 The left panels show the total variance of the winter average (December, January, February) 500mb height fields. The middle panels show that part of the total variance that is due to ENSO. The right panels show the ratio of the two variances (ENSO/Total). The top panels are from a reanalysis and the bottom panels are from atmospheric general circulation model (AGCM) simulations forced with observed sea surface temperatures. The results are computed for the period from 1950 to 1999, and plotted for the Northern Hemisphere polar cap to 20°N. The contour interval is 1000 (m²) in the left and middle panels, and 0.1 in the right panels (taken from Hoerling and Kumar 2002).

of large differences between the reanalyses in the various components of the atmospheric water cycle (*e.g.*, precipitation and evaporation). This problem can be alleviated to some extent by indirectly estimating the physical processes from other related quantities that are better constrained by the observations (*e.g.*, Sardeshmukh, 1993). Nigam *et al.* (2000) show, for example, that the heating obtained from a residual approach appears to be of sufficient quality to diagnose errors in the ENSO-heating distribution in a climate model simulation.

Another approach to addressing errors in the forcing is to focus directly on the adjustments made to the model forecast during the assimilation (*e.g.*, Schubert and Chang, 1996; Jeuken *et al.*, 1996; Rodwell and Palmer, 2007). These corrections can potentially provide substantial information about model limitations. Typically, the biases seen in fields, such as the monthly average temperature, are the result of complex interactions among small errors in different components of the model that grow over time. The challenge to modelers is to determine the individual potential sources of error, and ultimately to correct the inadequacies at the process level to improve long-term model behavior.

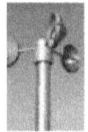

The development of a data assimilation system that provides unbiased estimates of the various physical processes inherent in the climate system is an important step in efforts to explain, or attribute, the causes of climate anomalies.

An important aspect of the corrections made during data assimilation is that they are applied frequently (typically every six hours), such that the impact of the adjustments can be seen before they can interact with the full suite of model processes. In other words, the corrections made during the course of data assimilation give a potentially direct method for identifying errors in the physical processes that create model biases (*e.g.*, Klinker and Sardeshmukh, 1992; Schubert and Chang, 1996; Kaas *et al.*, 1999; Danforth *et al.*, 2007; Rodwell and Palmer, 2007). They can also give insights into missing model physics such as dust-caused heating in the lower atmosphere (Alpert *et al.*, 1998), radiative heating in the stratosphere from volcanic eruptions (Andersen *et al.*, 2001), and impacts of land use changes (Kalnay and Cai, 2003)—processes not represented in the models used in the first reanalyses.

The development of a data assimilation system that provides unbiased estimates of the various physical processes inherent in the climate

system (*e.g.*, precipitation, evaporation, cloud formation) is an important step in efforts to explain, or attribute (Chapter 3), the causes of climate anomalies. Therefore, reanalyses allow scientists to go beyond merely documenting what happened. Scientists can, for example, examine the processes that maintain a large precipitation deficit in some region. Is the deficit maintained by local evaporative processes or by changes in the storm tracks that bring moisture to that region, or some combination of such factors? As described in Chapter 3, reanalysis data provide the first steps in a process of attribution (how well the causes of climate variability are understood) that involves detection and description of the anomalies, and an assessment of the important physical processes that contribute to their development. Ultimately, scientists seek answers to questions about the causes that cannot be addressed by reanalysis data alone. Going back to the previous example, how can the role of local evaporative changes and changes in the storm tracks be separated? Model experimentation is required, as described in Chapter 3: here too, reanalyses play an important role in validating the model behavior.

2.2.4 Outlook

There are a number of steps that can be taken to increase the value of reanalyses for identifying model deficiencies, including: improving our estimates of uncertainties in all reanalysis products, balancing budgets of key quantities (*e.g.*, heat, water vapor, energy) (Kanamitsu and Saha, 1996; see also the next Section), and reducing the false model response to the adjustments made to the background forecast by the insertion of observations (the so-called model spin-up or spin-down problem), especially when the adjustments involve water vapor and the various components of the hydrological cycle (Kanamitsu and Saha, 1996; Schubert and Chang, 1996; Jeuken *et al.*, 1996). For example, Annan *et al.* (2005) proposed an ensemble forecast approach to estimating model parameters. These, and other approaches, hold substantial promise for obtaining optimal estimates of uncertain model parameters from reanalyses, even for the current comprehensive climate models.

BOX 2.1: The Complementary Roles of Reanalysis and Free-Running Model Simulations in the Attribution Problem

Section 2.3 demonstrates the value of reanalysis for identifying and understanding climate variability. By providing best estimates of the circulation patterns and other weather elements, such as moisture transport, evaporation, precipitation, and cloudiness, which are present during observed extremes—estimates that are comprehensive and consistent over space and time—reanalysis offers a unique and profound contribution to the more general problem of attribution discussed in Chapter 3. Reanalyses are especially useful for providing a global picture of the prevailing anomalous circulation patterns such as those associated with a given drought. By studying reanalysis data, investigators can hypothesize linkages between the drought and climate anomalies in other parts of the world (e.g., anomalies in sea surface temperatures [SSTs]).

Reanalysis is one tool for addressing the problem. A drawback of reanalysis in this context is its inability to isolate causality—to demonstrate unequivocally that one climate feature (e.g., anomalous SSTs) causes another (e.g., drought). This drawback can extend to any set of direct observations of the atmosphere. Climate model simulations that are unconstrained by the assimilation of observational data are needed in order to isolate causality, Climate models can be forced in different ways to determine whether a certain forcing will cause the model to reproduce a climate anomaly of interest. For example, if an investigator suspects, perhaps based on an analysis of reanalysis data, that anomalous SSTs caused the severe drought in the southern Great Plains during the1950s, he or she could perform two simulations with a free-running climate model, one in which the 1950s SST anomalies are imposed, and one in which they are not. If only the first simulation reproduces the drought, the investigator has evidence to support the hypothesized role of the SSTs. An additional step would be to determine the cause of the SST anomalies, which would require further experiments with a comprehensive atmosphere/ocean/land model.

These free-running modeling studies have their own deficiencies, most importantly the potential lack of realism in the climate processes simulated by an unconstrained (non-reanalysis) modeling system. This suggests an important additional role of reanalysis in the attribution problem. Not only can the reanalysis data help in the formulation of hypotheses to be tested with a free-running climate model, but it can (and should) be used to verify that the free-running model is behaving realistically, *i.e.*, that the variations in circulation and other climate processes in the free-running model are consistent with what we have learned from reanalysis (see Section 2.2). Reanalysis and free-running model simulations are complementary tools for addressing the attribution problem, each with their own strengths and weaknesses. Only the unconstrained parts of a model can be used to address attribution (causality), implying the need for free-running models, but those unconstrained parts must be evaluated for realism, implying the need for reanalysis. Arguably, the best approach to the attribution problem is to use the reanalysis and free-running model approaches in tandem.

2.3. USING CURRENT REANALYSES TO IDENTIFY AND UNDERSTAND MAJOR SEASONAL-TO-DECADAL CLIMATE VARIATIONS

In this Section the strengths and weaknesses of current reanalyses for identifying and understanding climate variability are examined. This is an important step for addressing the more general issue of attribution, which was introduced in Chapter 1 and is addressed more fully in Chapter 3. Understanding the connections between reanalysis, models, and attribution is crucial for understanding the broader path towards attribution, as outlined in Chapter 1 (see Box 2.1).

2.3.1 Climate Variability

The climate system varies greatly over space and time. The variability of the atmosphere in particular encompasses common, individual weather events, and longer-term changes affecting global weather patterns that can result in regional droughts or wet periods (pluvials) lasting many years. A primary research goal is to understand the causes of these long-term climate variations and to develop models that enable scientists to predict them.

On subseasonal to decadal time scales there are a number of key recurring global-scale patterns of climate variability that have pronounced impacts on the North American climate (Table 2.3), including the Pacific/North American pattern (PNA), the Madden-Julian

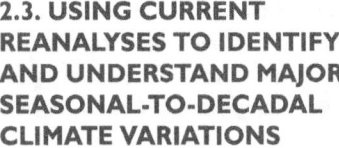

The variability of the atmosphere encompasses common, individual weather events and longer-term changes, affecting global weather patterns that can result in regional droughts or wet periods lasting many years.

Table 2.3 Characteristics of some of the leading modes of climate variability that are known to have a substantial impact on North American climate. The last column provides a subjective assessment of the quality of the atmospheric manifestations of these modes (and their impacts on regional climate) in current atmospheric reanalyses.

Phenomenon	Key reference	Time scale	Strength of link between atmosphere and ocean	Some impacts on North America	Consistency between atmospheric reanalyses
Pacific-North American (PNA) pattern	Wallace and Gutzler (1981)	Subseasonal-to-Seasonal	Weak to moderate	West coast storms	Good
Madden Julian Oscillation (MJO)	Madden and Julian (1994)	Approximately 30-60 days	Weak to moderate	Atlantic hurricanes	Fair to poor
North Atlantic Oscillation (NAO)	Hurrell et al. (2001)	Subseasonal-to-decadal	Moderate on long time scales	East coast winters	Good
Northern Annular Mode (NAM)	Thompson and Wallace (2000); Wallace (2000)	Subseasonal-to-decadal	Moderate on long time scales	East coast winters	Good to fair in stratosphere
El Niño-Southern Oscillation (ENSO)	Philander (1990)	Seasonal-to-inter-annual	Strong	Winter in west coast and southern tier of United States, Mexico, warm season regional droughts	Good to fair on longer time scales
Pacific Decadal Oscillation (PDO)	Zhang et al. (1997)	Decadal	Strong	Drought or pluvials over North America	Fair to poor
Atlantic Multi-decadal Oscillation (AMO)	Folland et al. (1986)	Decadal	Strong	Drought or pluvials over North America, Atlantic hurricanes	Fair to poor

Oscillation (MJO), the North Atlantic Oscillation (NAO) and the related Northern Annular Mode (NAM), the Quasi-Biennial Oscillation (QBO), El Niño-Southern Oscillation (ENSO), the Pacific Decadal Oscillation (PDO), and the Atlantic Multi-decadal Oscillation (AMO). These patterns, sometimes referred to as modes of climate variability or teleconnection patterns, can shift weather patterns and disrupt local climate features (*e.g.*, Gutzler *et al.*, 1988; Hurrell, 1996).

As discussed in the following Sections, the quality of the representation of these phenomena in reanalyses vary and depend on the time scales, locations, and physical processes relevant to each of these modes of variability. The last column in Table 2.3 gives the authors' expert assessment of the consistency of the atmospheric manifestations of these modes (and their impacts on regional climate) in current reanalyses based on such general considerations.

Figures 2.8 and 2.9 show examples of the connection between the PNA and NAO patterns and North American surface temperature and precipitation variations. The spatial correspondence between the reanalysis tropospheric circulation and the independently-derived surface patterns show the potential value of the reanalysis data for interpreting the relationships between changes in the climate modes and regional changes in surface temperature and precipitation.

During the positive phase of the PNA pattern, surface temperatures over western North America tend to be above average; this can be related to an unusually strong high pressure ridge over the region as well as transport of warm Pacific air poleward along the West Coast extending to Alaska. An upper-level trough centered over the Southeast United States and the associated intensified north to south flow over the center of the continent facilitates the southward transport of Arctic air that produces a tendency toward below normal temperatures over the Gulf Coast states. This same flow pattern is associated with transport of relatively dry polar air and a tendency to produce descending motions in the middle troposphere over the Missouri and Mississippi regions, both of which favor below normal precipitation, as observed. In contrast, the positive phase of the NAO pattern is accompanied by above average temperatures over the eastern United States and above average precipitation in the Ohio Valley. The reanalysis data of tropospheric circulation help to interpret this relationship as resulting from a northward-shifted westerly flow regime over the eastern United States and North Atlantic that inhibits cold air excursions while simultaneously facilitating increased moisture convergence into the region.

The above patterns arise mainly, but not exclusively, as manifestations of internal atmospheric variability; that is, they owe their existence largely to processes that are confined to the atmosphere such as various atmospheric instabilities and nonlinear processes (*e.g.,* Massacand and Davies, 2001; Cash and Lee, 2001; Feldstein, 2002, 2003; Straus and Shukla, 2002, and as discussed in Chapter 3). They are, however, also linked in varying degrees to processes external to the atmosphere such as interactions with the land surface and ocean variations. Understanding subseasonal-to-decadal climate variability requires that we understand the physical processes that produce these large-scale patterns, including how they interact with each other, and their interactions with the different climate system components (Chapter 3).

A key factor that limits scientists' ability to fully understand such long-term variability is the lack of long-term comprehensive and consistent observations of the climate system, including observations of the land and ocean, which are critical to understanding and predicting atmospheric variability over seasonal and longer time periods. Observations of each of these climate system components, while improving with increased satellite usage, are not yet sufficient for addressing climate problems. In order to adequately address seasonal and longer period of variability, the observations need to continuously cover many decades, span the globe, include all key climate parameters,

PNA Impact

Temperature

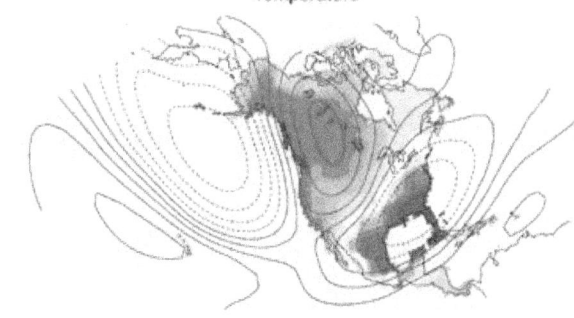

Precipitation

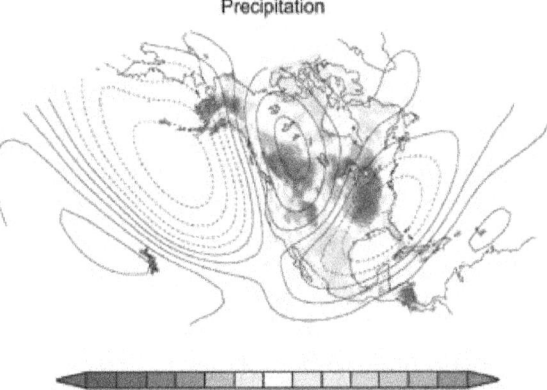

-0.7 -0.6 -0.5 -0.4 -0.3 -0.2 -0.1 0.1 0.2 0.3 0.4 0.5 0.6 0.7

Figure 2.8 The contours indicate the correlation between the wintertime PNA index (Wallace and Gutzler, 1981) and 500 millibar height field. The color shading indicates the correlations between the PNA index and the surface temperature (top panel) and the precipitation (bottom panel). The 500millibar height is from the NCEP/NCAR R1 reanalysis. The surface temperature and precipitation are from independent observational datasets. The correlations are based on seasonally-averaged data from 1951 to 2006. The contours of correlation give an indication of the direction of the mid-tropospheric winds, and the positions of the troughs and ridges.

and be consistent with our best physical understanding.

Among all components of the climate system, the atmospheric component possesses the most advanced observational capabilities. This system was developed primarily to support weather prediction, with major advances occurring first with the onset of a network of radiosondes in the 1950s and then with a near global observing system provided by satellite measurements beginning in the late 1970s. The present observing system is, however, still not fully adequate for many applications, and efforts continue to develop a true climate observing system that spans all climate system components and that provides continuity across space and time (GEOSS, 2005).

NAO Impact

Temperature

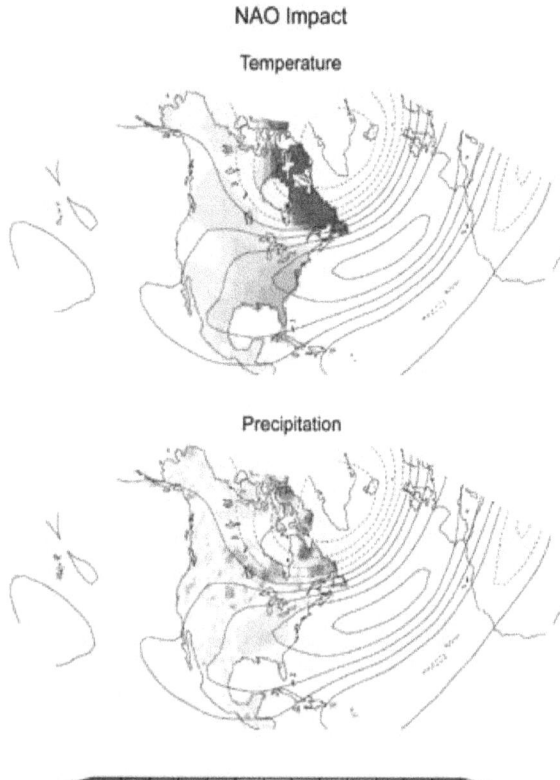

Precipitation

-0.7 -0.6 -0.5 -0.4 -0.3 -0.2 -0.1 0.1 0.2 0.3 0.4 0.5 0.6 0.7

Figure 2.9 The contours indicate the correlation between the wintertime NAO index (Wallace and Gutzler, 1981) and 500 millibar height field. The color shading indicates the correlations between the NAO index and the surface temperature (top panel) and the precipitation (bottom panel). The 500 millibar height is from the NCEP/NCAR R1 reanalysis. The surface temperature and precipitation are from independent observational datasets. The correlations are based on seasonally-averaged data from 1951 to 2006. The contours of correlation give an indication of the direction of the mid-tropospheric winds, and the positions of the troughs and ridges.

2.3.2 Reanalysis and Climate Variability
One of the most important insights of the last few decades regarding the existing observational record was that the investment in operational weather prediction could be leveraged by harnessing the prediction infrastructure (the global models and data assimilation methods for combining various observations) to develop a more consistent historical record of the atmosphere (Bengtsson and Shukla, 1988; Trenberth and Olson, 1988). This insight led to the development of several atmospheric climate reanalysis datasets (Schubert *et al.*, 1993; Kalnay *et al.*, 1996; Gibson *et al.*, 1997). These datasets provided the first comprehensive depictions of the global atmosphere that, in the case of the NCEP/NCAR reanalysis (Kalnay *et*

al., 1996), now span over 60 years. This Section summarizes how these and several follow-on reanalyses (Kanamitsu *et al.*, 2002; Uppala *et al.*, 2005; Onogi *et al.*, 2005; Mesinger *et al.*, 2006)[2] have contributed to an improved understanding of seasonal to decadal variability of climate (Table 2.1).

The reanalysis data provide the most comprehensive picture to date of the state of the atmosphere and its evolution. The reanalyses also provide estimates of the various physical processes, such as precipitation, cloud formation, and radiative fluxes, that are required to understand the processes by which climate evolves. As the utility of current reanalyses for identifying and understanding atmospheric variability is examined, the critical roles of the model in determining the quality of the reanalysis must be recognized, and the impact of the observing system inconsistencies in both space and time must also be appreciated. When assessing the utility of the reanalyses, the nature of the problem that is being addressed must also be considered. What is the time frame? How big is the area coverage? Does the problem involve the tropics or Southern Hemisphere, which tend to be less well observed, especially before the onset of satellite observations? To what extent are water vapor and clouds or links to the land surface or the ocean important? These are important considerations because data assimilation systems used for the first reanalyses evolved from numerical weather prediction needs; however, these systems did not place a high priority on modeling links to the land and ocean, which were considered to be of secondary importance to producing weather forecasts from a day to a week in advance.

The capacity of current reanalyses to describe and understand major seasonal-to-decadal climate variations is addressed in Sections 2.3.2.1, 2.3.2.2, and 2.3.2.3 by examining three key aspects of reanalyses: their spatial characteristics, their temporal characteristics, and their internal consistency and scope. Key examples are given of where reanalyses have contributed to the understanding of seasonal-to-decadal

---
[2] While not global, the North American Regional Reanalysis (NARR) has played an important role for studying regional climate variability. Two of its key strengths are the enhanced resolution, and the fact that precipitation observations were assimilated.

variability and where improvement is needed. This Product builds on the results of two major international workshops on reanalysis (WCRP, 1997; WCRP, 1999) by emphasizing studies that have appeared in the published literature since the last workshop.

Spatial characteristics

The globally complete spatial coverage provided by reanalyses, along with estimates of the physical processes that drive the atmosphere, has greatly facilitated diagnostic studies that attempt to identify the causes of large-scale atmospheric variability that have substantial impacts on North American weather and climate (*e.g.*, the NAO and PNA). Substantial improvements have been made in understanding the nature of both the NAO and PNA through studies using reanalysis products. Thompson and Wallace (2000), for example, provide a global perspective on the NAO, using reanalysis data to link it to the so-called Northern Hemisphere Annular Mode (NAM), noting the similarities of that mode to another annular mode in the Southern Hemisphere. Reanalysis data have also been used to link the variability of the NAO to that in the stratosphere in the sense that anomalies developing in the stratosphere propagate into the troposphere, suggesting a source of potential predictability over subseasonal time periods (*e.g.*, Baldwin and Dunkerton, 1999, 2001). Detailed studies made possible by reanalysis data have contributed to the understanding that both PNA and NAO modes of variability are fundamentally internal to the atmosphere,;that is, they would exist naturally in the atmosphere without any anthropogenic or other "external" forcing (*e.g.*, Massacand and Davies, 2001; Cash and Lee, 2001; Feldstein, 2002, 2003; Straus and Shukla, 2002; see also Chapter 3 on attribution). Straus and Shukla (2002) emphasized the differences between the PNA and a similar pattern of variability in the Pacific/North American region that is forced primarily as an atmospheric response to the tropical sea-surface temperature changes associated with ENSO.

Reanalysis data also allow in-depth evaluations of the physical processes and global connections of extreme regional climate events such as droughts or floods. For example, Mo *et al.* (1997), building on several earlier studies (*e.g.*, Trenberth and Branstator, 1992; Trenberth

and Guillemot, 1996), capitalized on the long record of the NCEP/NCAR global reanalyses to provide a detailed analysis of the atmospheric processes linked to floods and droughts over the central United States, including precursor events connected with large-scale wave propagation and changes in the Great Plains low level jet (LLJ). Liu *et al.* (1998) used reanalysis data in conjunction with a linear model to deduce the role of various physical and dynamical processes in the maintenance of the circulation anomalies associated with the 1988 drought and 1993 flood over the United States.

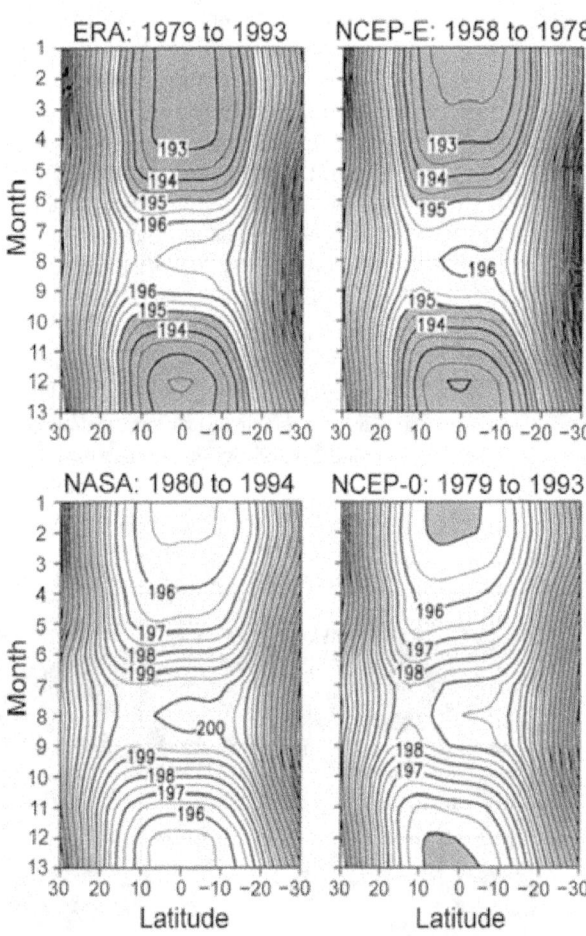

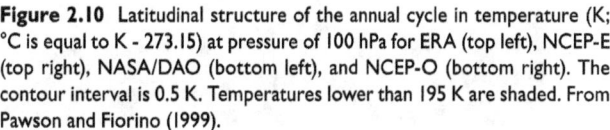

Figure 2.10 Latitudinal structure of the annual cycle in temperature (K; °C is equal to K - 273.15) at pressure of 100 hPa for ERA (top left), NCEP-E (top right), NASA/DAO (bottom left), and NCEP-O (bottom right). The contour interval is 0.5 K. Temperatures lower than 195 K are shaded. From Pawson and Fiorino (1999).

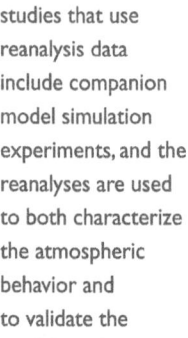

Many recent
studies that use
reanalysis data
include companion
model simulation
experiments, and the
reanalyses are used
to both characterize
the atmospheric
behavior and
to validate the
model results.

Process studies focused on North America have benefited from the high resolution and improved precipitation fields of the North American Regional Reanalysis (NARR). The studies examine, for example, the nature and role of the LLJ (*e.g.*, Weaver and Nigam, 2008), land-atmosphere interactions (*e.g.*, Luo *et al.*, 2007), and efforts to validate precipitation processes in global climate models (*e.g.*, Lee *et al.*, 2007). These studies highlight the leading role of reanalysis data in the diagnostic evaluation of large-scale climate variability and of the physical mechanisms that produce high impact regional climate anomalies.

While reanalysis data have played a fundamental role in diagnostic studies of the leading middle- and high-latitude variability and of regional climate anomalies, there are inadequacies in the stratosphere—a region of the atmosphere particularly poorly resolved in initial reanalysis systems (*e.g.*, Pawson and Fiorino, 1998a,b, 1999; Santer *et al.*, 2003), but better represented in more recent reanalyses, such as the ERA-40 (Santer *et al.*, 2004). Figure 2.10 shows an example of the substantial differences between the reanalyses that occur in the tropical stratosphere even in such a basic feature as the annual cycle of temperature.

Another area of concern is in polar regions where the reanalysis models have limitations in both the numerical representation and the modeling of physical processes (*e.g.*, Walsh and Chapman, 1998; Cullather *et al.*, 2000; Bromwich and Wang, 2005; Bromwich *et al.*, 2007). In particular, reanalyses have been inadequate in the modeled polar cloud properties

and associated radiative fluxes (*e.g.*, Serreze *et al.*, 1998).

Variations in tropical sea surface temperatures (SST), especially those associated with ENSO, are a major contributor to climate variability over North America on interannual time scales (*e.g.*, Trenberth *et al.*, 1998). Recent studies that use reanalysis data have contributed to important new insights on the links between tropical Pacific SST variability and extratropical circulation (*e.g.*, Sardeshmukh *et al.*, 2000; Hoerling and Kumar, 2002; DeWeaver and Nigam, 2002), the global extent of the ENSO response (*e.g.*, Mo, 2000; Trenberth and Caron, 2000), and its impact on weather (*e.g.*, Compo *et al.*, 2001; Gulev *et al.*, 2001; Hodges *et al.*, 2003; Raible, 2007; Schubert *et al.*, 2008). Many of these studies include companion model simulation experiments, and the reanalyses are used to both characterize the atmospheric behavior and to validate the model results. This is an important advance in climate diagnosis resulting from increased confidence in climate models, and it represents an important synergy between reanalysis and the attribution studies discussed in Chapter 3.

While the reanalyses are useful in many respects for addressing the problem of tropical/extratropical connections, there are limitations in representing tropical precipitation, clouds, and other aspects of the hydrological cycle (*e.g.*, Newman *et al.*, 2000). The Madden-Julian Oscillation is an example of a phenomenon in which the interaction between the circulation and tropical heating is fundamental to its structure and evolution (*e.g.*, Lin *et al.*, 2004)—an interaction that has not yet been well represented in climate models. Current reanalysis products are inadequate for validating models because those aspects of the MJO that appear to be important for proper simulation (*e.g.*, the vertical distribution of heating) are poorly constrained by observations and are therefore highly dependent on the models used in the assimilation systems (*e.g.*, Tian *et al.*, 2006). Indirect (residual) approaches to estimate the tropical forcing from reanalyses, however, can be useful, reflecting the greater confidence placed in the estimates of certain aspects of the large-scale tropical circulation (Newman *et al.*, 2000; Nigam *et al.*, 2000).

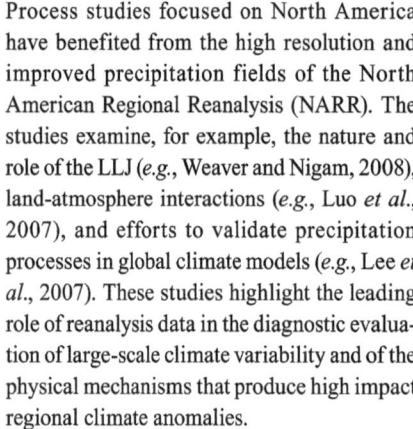

While the NAO, PNA and ENSO phenomena influence subseasonal-to-interannual climate variability, there is evidence that these modes also may vary over periods of decades or longer. Understanding that behavior, as well as other decadal-scale modes of variability such as the Pacific Decadal Oscillation and the Atlantic Multi-decadal Oscillation, require datasets that are consistent over many decades.

Temporal characteristics

The observing system over the last century varies greatly over time. Prior to the mid-twentieth century, the observing system was primarily surface-based and limited to land areas and ship reports, although some higher observations (*e.g.*, wind measurements from pilot balloons) have been made routinely since the early twentieth century (*e.g.*, Brönnimann *et al.*, 2005). An upper-air radiosonde network of observations was initiated in the late 1940s but was primarily confined to land areas, and Northern Hemisphere midlatitudes in particular. A truly global observing system arose with the onset of satellite observations in the 1970s, with numerous changes made to the observing system as new satellites were launched with updated and more capable sensors, and older systems were discontinued (Figure 2.2). The changes in the observing system, together with improved sensors and the aging and degrading of existing sensors, makes combining all available observations into a consistent long-term global climate record a major challenge. Figure 2.11 provides an overview of the number of observations made at all latitudes from 1946 to 1998 that were available to the NCEP/NCAR reanalysis (Kistler *et al.*, 2001). These changes, especially the onset of satellite observations, have impacted the reanalysis fields, often making it difficult to separate true climate variations from artificial changes associated with the evolving observing system.

The changes in the observing system have impacted the ability to study variability on interannual and longer time periods—the time scales at which changes to the observing system also tend to occur (*e.g.*, Basist and Chelliah, 1997; Chelliah and Ropelewski, 2000; Kistler *et al.*, 2001; Trenberth *et al.*, 2001; Kinter *et al.*, 2004). The impact can be complicated, involving interactions and feedbacks with the

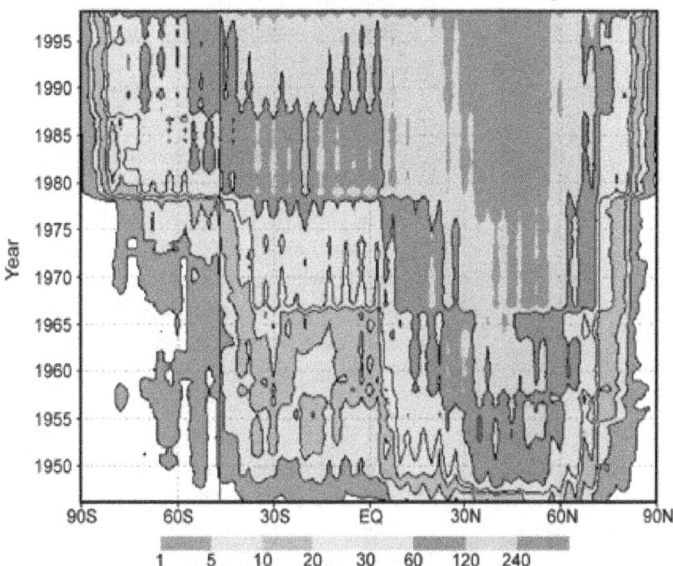

Number of Observations / 2.5deg

Figure 2.11 Zonal average number of all types of observations available to the NCEP/NCAR reanalysis per 2.5° latitude-longitude box per month from 1946 to 1998. A 12-month running average has been applied. From Kistler *et al.* (2001).

assimilation schemes. For example, Trenberth *et al.* (2001) show how discontinuities in tropical temperature and moisture can be traced to the bias correction of satellite radiances in the ECMWF (ERA-15) reanalyses. Changes in conventional radiosonde observations can also have impacts. For example, the Quasi-Biennial Oscillation, while clearly evident throughout the record of the NCEP/NCAR reanalysis, shows substantial secular changes in amplitude that are apparently the result of changes in the availability of tropical wind observations (Kistler *et al.*, 2001). The major change in the observing system associated with the onset of satellite data in the 1970s represents a particularly difficult and important problem because it coincides with the time of a major climate shift associated with the Pacific Decadal Oscillation (*e.g.*, Pawson and Fiorino, 1999; Trenberth and Caron, 2000; Chelliah and Bell, 2004).

Despite these problems, reanalysis data can be valuable in understanding long-term atmospheric variability, particularly if used in conjunction with other independent observations. For example, Barlow *et al.* (2001) used NCEP/NCAR reanalyses of winds and stream function for the period 1958 to 1993, in conjunction with independent sea surface temperature, streamflow, precipitation, and other data to identify

> The changes in the observing system, together with improved sensors and the aging and degrading of existing sensors, makes combining all available observations into a consistent long-term global climate record a major challenge.

three leading modes of SST variability affecting long-term drought over the United States.

In general, the quality of reanalysis tends to be best at weather time scales of a day to about a week, and degrades over both shorter and longer periods of time. The changes in quality reflect both the changes in the observing system and the ability of the model to simulate the variability at the different lengths of time. For time periods of less than a day, there are several factors that degrade the quality of the analysis. These include an observing system that does not fully resolve variations shorter than one day, and deficiencies in model's representation of the diurnal cycle (*e.g.*, Higgins *et al.*, 1996; Betts *et al.*, 1998a). This issue also contributes to errors in our estimates of seasonal and longer time averages of reanalysis quantities. It is not surprising that the quality is best for the weather time scales (*e.g.*, Beljaars *et al.*, 2006), since the analysis systems and models used thus far for atmospheric reanalyses were developed for global numerical weather prediction.

There are also important connections between the atmosphere and the land and ocean systems on seasonal and longer periods of time that can limit reanalysis quality if they are not fully understood. The assimilation systems for the land and ocean components are considerably less developed than for the atmosphere (discussed further in Section 2.5). The connection between the atmosphere and the ocean in the current generation of atmospheric reanalyses is made by specifying sea surface temperatures from reconstructions of historical observations;

the land is represented in a simplified form, which can also contribute to limitations in representing the diurnal cycle because the cycle is interconnected with the land surface (*e.g.*, Betts *et al.*, 1998b).

Model errors can have particularly large impacts on quantities linked to the hydrological cycle, such as atmospheric water vapor (*e.g.*, Trenberth *et al.*, 2005) and major tropical circulations (*e.g.*, the Hadley Cell) that are relevant to understanding climate variations and change (Mitas and Clement, 2006). Any bias in the model can exacerbate false climate signals associated with a changing observing system, for example, a model that consistently produces conditions that are too dry in the lower atmosphere. Such a model may give a realistic tropical precipitation condition when there are few moisture observations available to constrain the model, but that same model might produce unrealistic rainfall for the satellite era when it is confronted with large amounts of water vapor information that is inconsistent with the model's average water vapor distribution (Figure 2.5).

The impacts of the changing observing systems on current reanalysis products indicate these changes have not yet been accounted for. To date, all available observations have been used in order to maximize the accuracy of the reanalysis products at any given time, but efforts to develop approaches that would reduce the inconsistencies over long time periods in the reanalysis products have been limited. This issue has been recognized, and efforts are currently underway to carry out reanalyses with a subset of the full observing systems to try to minimize the changes over time (*e.g.*, Compo *et al.*, 2006), as well as to conduct other observing system sensitivity experiments that could help to understand, if not reduce, the impacts (*e.g.*, Bengtsson *et al.*, 2004b,c; Dee, 2005; Kanamitsu and Hwang, 2006). Model bias correction techniques (*e.g.*, Dee and da Silva, 1998; Chepurin *et al.*, 2005; Danforth *et al.*, 2007), improvements to our models (Grassl, 2000; Randall, 2000), and improvements to historical observations including data mining, improved quality control and further cross calibration and bias correction of observations (Schubert *et al.*, 2006) may also help to reduce the impacts from the changing observing system.

> There are important connections between the atmosphere and the land and ocean systems on seasonal and longer periods of time that can limit reanalysis quality if they are not fully understood.

Internal consistency and scope

An advantage of the reanalysis products mentioned earlier involves the role of the model in providing internal consistency, meaning that the model enforces certain dynamical balances that are known to exist in the atmosphere, such as the tendency for the atmosphere to be in geostrophic balance (an approximate balance of the Coriolis and pressure gradient forces) in the midlatitudes. One important implication is that the different state variables (the quantities that define the state of the atmosphere—*e.g.*, the winds, temperature, and pressure) depend strongly on one other. That such constraints are satisfied in the reanalysis products is important for many studies that attempt to understand the physical processes or forcing mechanisms by which the atmosphere evolves (*e.g.*, the various patterns of variability mentioned above).

A fundamental advantage of model-based reanalysis products over single variable analyses of, for instance, temperature or water vapor observations, is that reanalysis products provide a comprehensive, globally complete picture of the atmosphere at any given time, together with the various forcings that determine how the atmosphere evolves over time. In principle it is possible to diagnose all aspects of how the climate system has evolved over the time period covered by the reanalyses; however, the results depend on the quality of the model as well as model characteristics and observational errors used in the reanalysis. As mentioned earlier, the models used in the current generation of reanalyses were largely developed for midlatitude numerical weather prediction and have known limitations, especially in various components of the hydrological cycle (clouds, precipitation, evaporation) that are necessary for understanding such important phenomena as monsoons, droughts, and various tropical phenomena.

Given that models are imperfect, can model-based reanalysis products be used to validate model simulations (see also Section 2.2)? For example, by forcing models with the historical record of observed sea surface temperatures, can some of the major precipitation anomalies that occurred over the last century be accurately reproduced (*e.g.*, Hoerling and Kumar, 2003; Schubert *et al.*, 2004; Seager *et al.*, 2005; Chapter 3)? As these simulations are examined

for clues about how the climate system operates, there is an increasing need to validate the physical processes that produce the regional climate anomalies (*e.g.*, drought in the Great Plains of the United States). There is a question as to whether the reanalyses used in the validations are themselves compromised by model errors. However, evidence is growing that, at least in regions with relatively good data coverage, the reanalyses can be used to identify fundamental errors in the model forcing of hydrological climate anomalies (*e.g.*, Ruiz-Barradas and Nigam, 2005).

On global scales, the limitations in the assimilation models are shown as biases in, for example, monthly averaged heat and moisture budgets, introducing uncertainties in the physical processes that contribute to them (*e.g.*, Trenberth and Guillemot, 1998; Trenberth *et al.*, 2001; Kistler *et al.*, 2001). There has been success in looking at variability of the energy budgets associated with some of the major climate variations such as ENSO (*e.g.* Trenberth *et al.*, 2002a); however, inconsistencies in certain budgets (especially the atmospheric energy transports) limit their usefulness for estimating overall surface fluxes (Trenberth and Caron, 2001)—quantities that are important for linking the atmosphere and the ocean, as well as the atmosphere and land surface. Limitations in model-estimated clouds (and especially short wave radiation) appear to be a primary source of the problems in model fluxes both at the surface and at the top of the atmosphere (*e.g.*, Shinoda *et al.*, 1999). Figure 2.12 shows an ex-

> A fundamental advantage of model-based reanalysis products is that reanalysis products provide a comprehensive, globally complete picture of the atmosphere at any given time, together with the various forcings that determine how the atmosphere evolves over time.

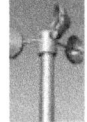

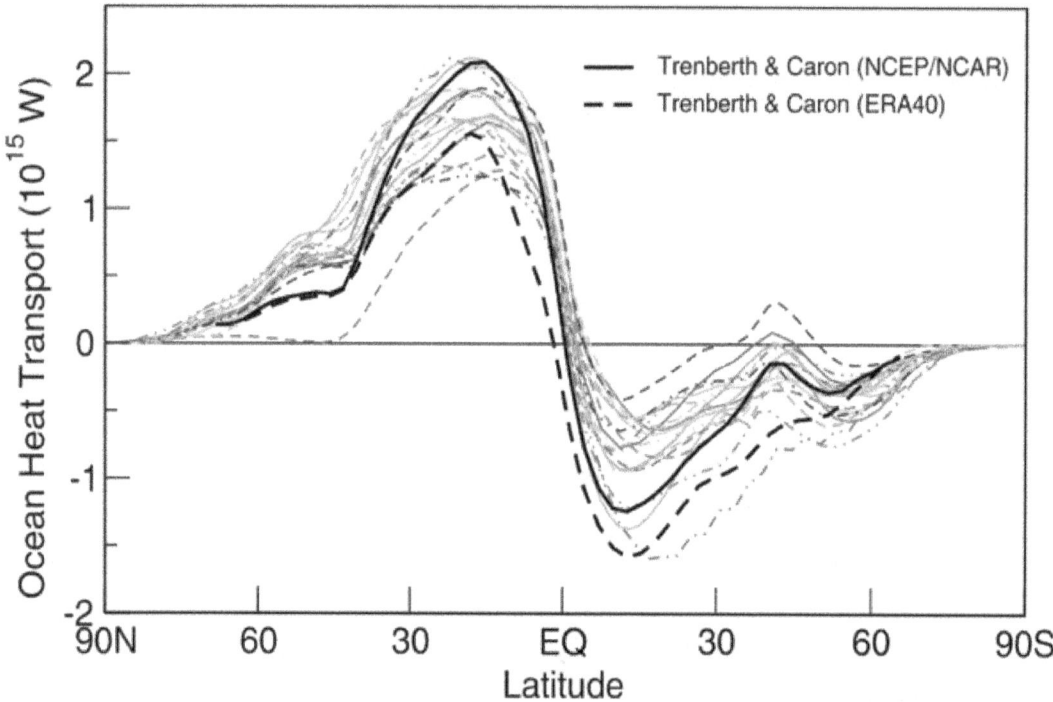

Figure 2.12 Annual mean, zonally-averaged oceanic heat transport implied by net heat flux imbalances at the sea surface, under an assumption of negligible changes in oceanic heat content. The observational based estimate, taken from Trenberth and Caron (2001) for the period February 1985 to April 1989, originates from reanalysis products from NCEP/NCAR (Kalnay et al., 1996) and European Centre for Medium Range Weather Forecasts 40-year reanalysis (ERA40; Uppala et al., 2005). The model averages are derived from the years 1980 to 1999 in the twentieth century simulations in the Multi-Model Dataset at the Program for Climate Model Diagnosis and Intercomparison (PCMDI). The legend identifying individual models appears in Figure 8.4 of the IPCC Fourth Assessment Report (IPCC, 2007).

The climate of a region is defined by statistical properties of the climate system evaluated over an extended period of time, typically over decades or longer.

ample of implied ocean heat transport estimates provided by two different reanalyses and how they compare with the values obtained from a number of different coupled atmosphere-ocean model simulations.

Current atmospheric reanalysis models do not satisfactorily represent interactions with other important components of the climate system (ocean, land surface, cryosphere). As a result, various surface fluxes (*e.g.*, precipitation, evaporation, radiation) at the interfaces between the land and atmosphere, cryosphere and atmosphere, and the ocean and atmosphere, are generally inconsistent with one other and therefore limit the ability to fully understand the forcings and interactions of the climate system (*e.g.*, Trenberth *et al.*, 2001). While there are important stand-alone land (*e.g.*, Reichle and Koster, 2005) and ocean (*e.g.*, Carton *et al.*, 2000) reanalysis efforts currently either in development or underway (Section 2.5), the long-term goal is a fully coupled climate reanalysis system (Tribbia *et al.*, 2003).

2.4 CLIMATE TRENDS IN SURFACE TEMPERATURE AND PRECIPITATION DERIVED FROM REANALYSES *VERSUS* FROM INDEPENDENT DATA

The climate of a region is defined by statistical properties of the climate system (*e.g.*, averages, variances, and other statistical measures) evaluated over an extended period of time, typically over decades or longer. If these underlying statistical values do not change with time, the climate would be referred to as "stationary". For example, in a stationary climate the average monthly rainfall in a specific region during the twentieth century, for instance, would be the same as that in the nineteenth, eighteenth, or any other century (within statistical sampling errors). Climate, however, is non-stationary; the underlying averages (and other statistical measures) do change over time. The climate system varies through ice ages and warmer periods with a timescale of about 100,000 years (Hays *et al.*, 1976). The "Little Ice Age" in the

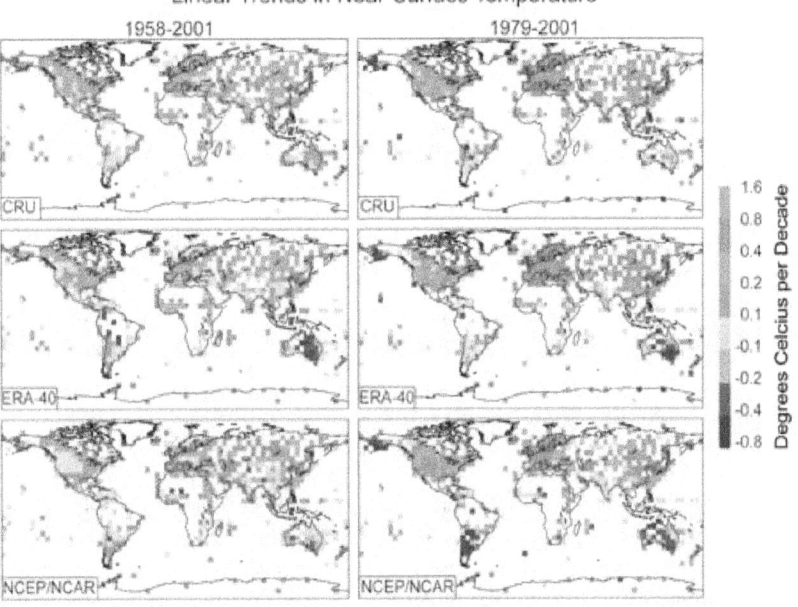

Linear Trends in Near-Surface Temperature

Figure 2.13 Calculated trends in near-surface (2 meter) temperature from an observational dataset (top), the ERA-40 reanalysis (middle), and the NCEP/NCAR reanalysis (bottom). Reproduced from Simmons *et al.* (2004).

fifteenth to nineteenth centuries (Bradley *et al.*, 2003) is an example of a natural climate variation (non-stationarity) with a much shorter timescale of a few centuries. Humans may be affecting climate even more quickly through their impact on atmospheric greenhouse gases (Hansen *et al.*, 1981).

The search for trends in climatic data is an attempt to quantify the non-stationarity of climate, as reflected in changes in long-term average climate values. There are various methods for accomplishing this task (see CCSP, 2006: Appendix A for a more detailed discussion). Perhaps the most common approach to calculating a trend from a multiple decade dataset is to plot the data value of interest (*e.g.*, rainfall) against the year of measurement. A line is fit through the points using standard regression techniques, and the resulting slope of the line is a measure of the climatic trend. A positive slope, for example, suggests that the "underlying climatic average" of rainfall is increasing with time over the period of interest. Such a trend calculation is limited by the overall noisiness of the data and by the length of the record considered.

2.4.1 Trend Comparisons: Reanalyses *Versus* Independent Measurements

Reanalysis datasets now span several decades, as do various ground-based and space-based measurement datasets. Trends can be computed from both. A natural question is: How well do the trends computed from the reanalysis data agree with those computed from independent datasets? This question has been addressed in many independent studies. Calculating trends is one method for assessing the adequacy of reanalysis data for evaluating climate trends. The focus here is on trends in two particular variables: surface temperature at a height of two meters, referred to here as T_{2m}, and precipitation.

Simmons *et al.* (2004) provide the most comprehensive evaluation to date of reanalysis-based trends in surface temperature, T_{2m}. Figure 2.13, reproduced from that work which uses linear regression techniques, shows comparison of T_{2m} from observations (the CRUTEM2v dataset of Jones and Moberg, 2003), with two reanalyses (ERA-40 and NCEP/NCAR).

The period from 1958 to 2001 (left) and from 1979 to 2001 (right) were considered. All three

Calculating trends is one method for assessing the adequacy of reanalysis data for evaluating climate trends.

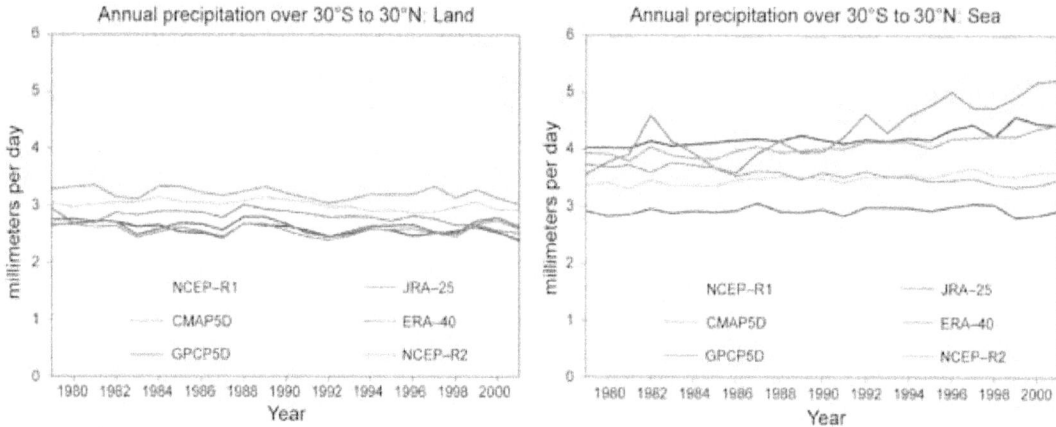

Figure 2.14 Annual tropical precipitation over land (left) and ocean (right) from four reanalyses (NCEP-R1, NCEP-R2, JRA-25, and ERA-40) and from two observational datasets (CMAP5D and GCPC5D). Reprinted from Takahashi *et al.* (2006).

datasets show generally positive trends. The reanalyses-based trends, however, are generally smaller, particularly for the longer time period. The average trend for 1958 to 2001 in the Northern Hemisphere, is 0.19°C per decade for the observations, 0.13°C for ERA-40, and 0.14°C for NCEP/NCAR. For the shorter and more recent period, the Northern Hemisphere averages are 0.30°C for the observations, 0.27°C for ERA-40, and 0.19°C for NCEP/NCAR. Simmons *et al.* (2004) consider the latter result for ERA-40 to be particularly encouraging because

"the agreement is to within about 10 percent in the rate of warming of the land areas of the Northern Hemisphere since the late 1970s". Stendel *et al.* (2000) note that for the ERA-15 reanalysis, which covers 1979 to 1993 using an earlier version of the modeling system, the trend in T_{2m} over North America and Eurasia is too small by 0.14°C per decade, relative to observations. Thus, the later ERA-40 reanalysis appears to improve significantly over the earlier ERA-15 reanalysis for T_{2m} temperature trends. Figure 2.13 shows that the performance of

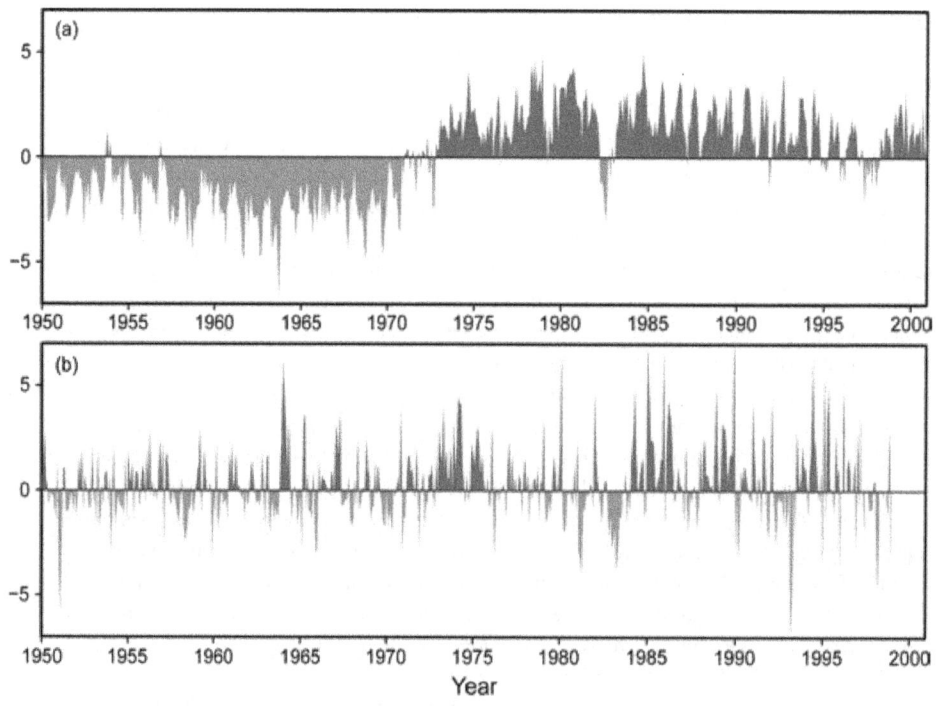

Figure 2.15 Precipitation averaged over 10°S-equator, 55°-45°W with respect to time, from (a) the NCAR/NCEP reanalysis, and (b) from an observational precipitation dataset. Reprinted from Kinter *et al.* (2004).

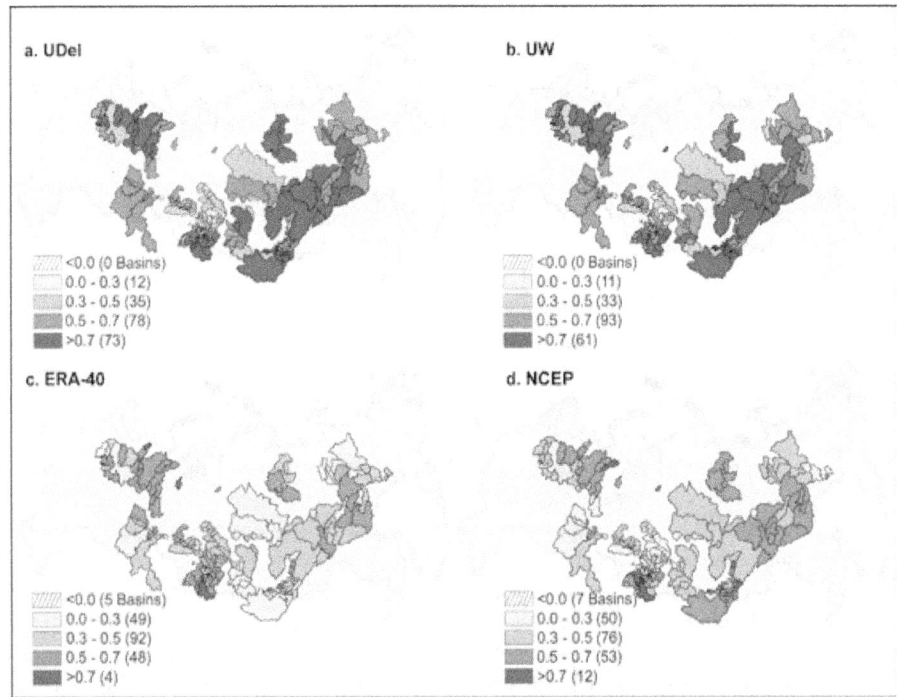

Figure 2.16 Identification of northern Asia river basins for which the computed precipitation trend is positive (a wetting trend) or negative (a drying trend), for four datasets: (top left) a dataset based on ground-based measurements of rainfall; (top right) a modified (improved) version of the first dataset; (bottom left) ERA-40 reanalysis; and (bottom right) NCEP/NCAR reanalysis. From Pavelsky and Smith (2006).

ERA-40 and NCEP/NCAR varies with region, with some clear areas of large discrepancies that most likely represent reanalysis errors. Both reanalyses underestimate trends in India and Australia. The NCEP/NCAR reanalysis in particular does not adequately reproduce trends in southern South America, a problem also noted by Rusticucci and Kousky (2002).

A similarly comprehensive evaluation of precipitation trends from reanalyses has not been published. Takahashi *et al.* (2006), however, do summarize the trends in total tropical (30°S to 30°N) precipitation over the period of 1979 to 2001 (Figure 2.14) based on two sets of observational data and four reanalyses.

The biggest discrepancy between the observations and reanalyses is the large positive trend over the ocean for the ERA-40 reanalyses and the smaller but still positive trends for the other reanalyses, trends that are not found in the observations. Similarly, Chen and Bosilovich (2007) show that the reanalyses indicate a positive precipitation trend in the 1990s when global precipitation totals are considered, whereas

observational datasets do not. By starting in 1979, the tropical analysis of Takahashi *et al.* (2006) misses a problem discovered by Kinter *et al.* (2004), who demonstrate a false precipitation trend produced by the NCEP/NCAR reanalysis in equatorial Brazil. As shown in Figure 2.15, the NCEP/NCAR reanalysis produces a strong, apparently unrealistic, increase in rainfall starting in about 1973, and thus, an unrealistic wetting trend.

Pohlmann and Greatbatch (2006) found that the NCEP/NCAR reanalysis greatly overestimates precipitation in northern Africa before the late 1960s, resulting in an unrealistic drying trend. Pavelsky and Smith (2006), in an analysis of river discharge to the Arctic Ocean, compared precipitation trends in the ERA-40 and NCEP/NCAR reanalyses with those from ground-based observations and found the reanalyses trends to be much too large, particularly for ERA-40. Figure 2.16 qualitatively summarizes these results.

River basins with an increasing precipitation trend and those with a decreasing precipita-

Compared with temperature trends, reanalysis-based precipitation trends appear to be less consistent with those calculated from observational datasets.

tion trend are identified for each dataset. For ERA-40, the vast majority of basins show an unrealistic (relative to ground observations) wetting trend.

Reanalyses that rely solely on atmospheric data may miss real trends in surface temperature that are associated with land usage, such as urbanization, cropland conversion, changing irrigation practices, and other land use changes.

2.4.2 Factors Complicating the Calculation of Trend
The previous studies indicate that observed temperature trends are captured to a large extent by the reanalyses, particularly in the latter part of the record, although some area trends (*e.g.*, Australia) have been more difficult to reproduce. Compared with temperature trends, reanalysis-based precipitation trends appear to be less consistent with those calculated from observational datasets. As described below, many studies have identified sources for errors with the reanalyses that at least partly explain these inadequacies; however, trends produced from the observational datasets are also subject to errors for several reasons (see CCSP, 2006, and discussed below), such that the true inadequacies of the reanalyses-based trends cannot be fully measured.

First, and perhaps most importantly, a false trend in the reanalysis data may result from a change in the observations being assimilated. In particular, with the onset of satellite data in the late 1970s, global-scale observations of highly variable quality increased dramatically. Consider a model that tends to "run cold" (has a negative temperature bias) when not constrained by data. If this model is used to perform a reanalysis of the last 50 years but by necessity only ingests satellite data from the late 1970s onward, then the first half of the reanalysis will be biased cold relative to the second half, leading to an artificial positive temperature trend (Figure 2.17).

Bengtsson *et al.* (2004a) use this reasoning to explain an apparently false trend in lower troposphere temperature (not surface temperature) produced by the ERA-40 reanalysis. Kalnay *et al.* (2006), when computing trends in surface air temperature from the NCEP/NCAR reanalysis, separate the 40-year reanalysis period into a pre-satellite and post-satellite period to avoid such issues. However, reanalyses can also be affected by non-satellite measurement system changes. Betts *et al.* (2005) note in reference to the surface temperature bias over Brazil that "the Brazilian surface synoptic data are not included [in the ERA-40 reanalysis] before 1967, and with its introduction, there is a marked shift in ERA-40 from a warm to a cool bias in two meter temperature".

Reanalyses that rely solely on atmospheric data may miss real trends in surface temperature that are associated with land usage, such as urbanization, cropland conversion, changing irrigation practices, and other land use changes (Pielke *et al.*, 1999; Kalnay *et al.*, 2006). The ERA-40 reanalysis, which assimilates some station-based air temperature measurements made at the surface, is less affected by this issue than the NCEP/NCAR reanalysis, which does not. This difference in station data assimilation may partially explain why ERA-40 reanalysis performs better compared with NCEP/NCAR reanalysis, as shown in Figure 2.13 (Simmons *et al.*, 2004).

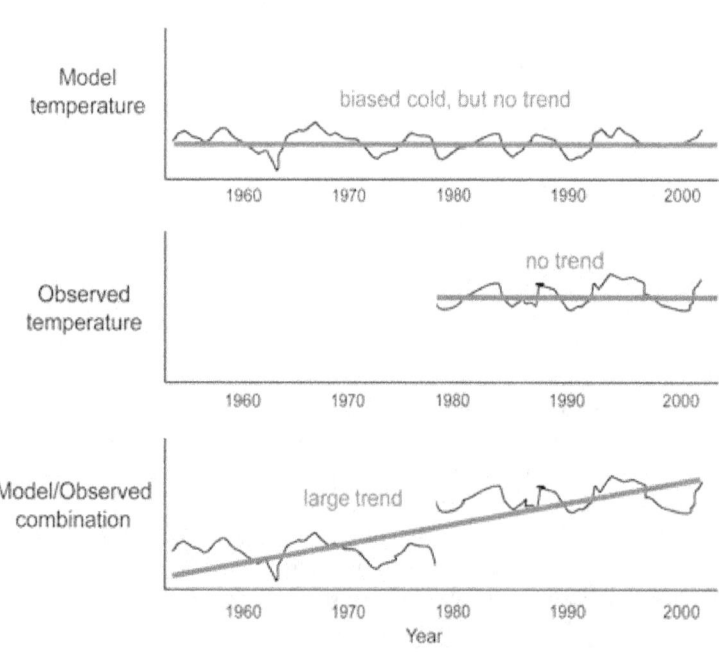

Figure 2.17 Idealized example showing how the correction of biased model data with observational data during only one part of a reanalysis period, from 1979 onward, can lead to a spurious temporal trend in the reanalysis product.

As mentioned above, calculating trends from observational datasets also involves errors, and introduces additional uncertainties when compared with reanalysis products, in which values are provided on regular grids. An important and challenging issue is estimating the appropriate grid-cell averaged temperature and precipitation values from point observations so that they can be directly compared with reanalysis products. Errors in representation may play an important role. For example, rainfall at one observation point may not be representative of rainfall over the corresponding model grid cell, which represents an area-average value. Rainfall measurements are often sparse and distributed non-randomly, *for example*, in the mountainous western United States, much of the precipitation falls as snow at high elevations, while most direct measurements are taken in cities and airports located at much lower elevations, and are therefore not representative of total precipitation in that region. Simmons *et al.* (2004) note that the gridded observational values along coastlines reflect mostly land-based measurements, whereas reanalysis values for coastal grid cells reflect a mixture of ocean and land conditions. Also, producing a gridded data value from multiple stations within the cell can lead to significant problems for trend estimation because the contributing stations may have different record lengths and other inhomogeneities over space and time (Hamlet and Lettenmaier, 2005). Jones *et al.* (1999) note that urban development over time at a particular sensor location can produce a positive temperature trend at the sensor that is real, but is likely unrepresentative of the large grid cell that contains it.

Observational datasets that span multiple decades are also subject to changes in measurement systems. Takahashi *et al.* (2006) suggest that the use of a new satellite data product (introduced in 1987) in an observational precipitation dataset led to a change in the character of the data. Kalnay *et al.* (2006) found an artificial trend in observational temperature data induced by changes in measurement time-of-day, measurement location, and thermometer type. Jones *et al.* (1999) discuss the need to adjust or omit station data as necessary to ensure a minimal impact of such changes before computing trends.

Figure 2.18 shows the uncertainty inherent in trend computations from various observational datasets, and compared with NCSEP/NCAR reanalysis.

The top six maps show the annual temperature trends across regions over the continental United States, as computed from six different observational datasets from 1951 to 2006, and the bottom map shows the trend computed from the NCEP/NCAR reanalysis. Of the seven maps, the reanalysis-derived map is clearly different from the other maps; the six observations-based maps all show a warming trend in all regions except the South, whereas

> Observational datasets that span multiple decades are subject to changes in measurement systems.

Annual Temperature Trends: 1951-2006

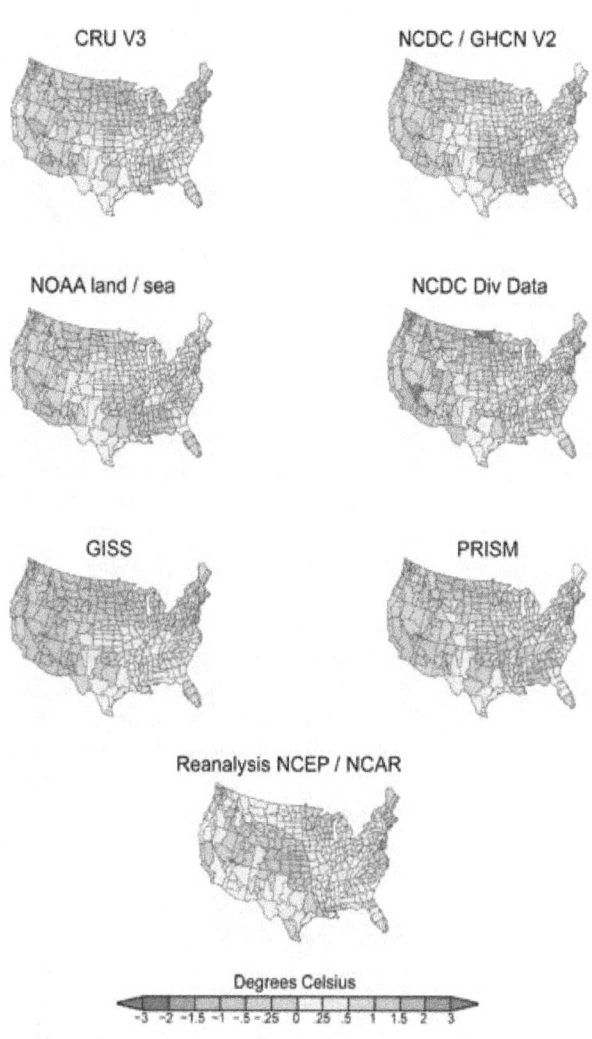

Figure 2.18 Annual temperature trends across the continental United States, as determined with six observational datasets and the NCEP/ NCAR reanalysis (M. Hoerling, personal communication).

the reanalysis shows a general warming in the South and cooling toward the West. However, the six observations-based maps do not fully agree with one another. For example, the area of cooling in the South is smaller in the GISS and CRU datasets than in the National Climatic Data Center (NCDC)/Global Historical Climatology Network (GHCN) dataset. The NCDC climate division data show relatively high temperature trends in the West. These maps illustrate the fact that there is no perfect "truth" against which to evaluate the reanalysis-based trends.

There are also other sources of uncertainty for both observations-based trends and reanalysis-based trends. The mathematical algorithm used to compute trends is important. Jones (1994a) uses the linear regression approach and the "robust trend method" of Hoaglin *et al.* (1983), thereby computing two similar, but not identical, sets of trend values from the same dataset. Also, part of the trend estimation problem is determining whether a computed trend is real, that is, the degree to which the trend is unlikely to be the result of statistical sampling variations. Groisman *et al.* (2004) describe a procedure they used to determine the statistical significance of computed trends, which can help alleviate this problem. Even if all surface temperature data were perfect and the trend estimation technique was not an issue, the time period chosen for computing a trend can result in sampling variations, depending, for example, on the relationship to transient events such as ENSO or volcanoes (Jones, 1994b).

2.4.3 Outlook
While limitations hamper the accurate estimation of trends from either reanalyses or observational datasets, it is the authors' assessment that it is likely that most of the trend differences shown in Figures 2.13 to 2.16 are related to limitations of the model-based reanalyses. Datasets that originate directly from surface and/or satellite observations, such as surface air temperature, precipitation, and atmospheric water vapor, will continue, at least for the near-term, to be the main tool for quantifying decadal and long-term climate changes. The observations-based trends are likely to be more reliable, in part because the relevant limitations in the observational data are better

A proper analysis of the mechanisms of climate trends requires substantial data, and only a reanalysis provides self-consistent datasets that are complete in space and time over several decades.

known and can, to a degree, be accounted for prior to trend estimation. This is less the case for existing reanalyses, which were not optimized for trend detection. Bengtsson *et al.* (2004a), examining various reanalysis products (though not surface temperature or precipitation), find that "there is a great deal of uncertainty in the calculation of trends from present reanalyses". Reanalysis-based precipitation (for ERA-40 and NCAR/NCEP) and surface air temperature (for NCAR/NCEP) are derived solely from the models (*i.e.*, precipitation and surface temperature observations are not assimilated). Therefore, these fields are subject to inadequacies in model parameterization. The North American Regional Reanalysis is an important example of a reanalysis project that did employ the assimilation of observed precipitation data (Mesinger *et al.*, 2006), producing, as a result, more realistic precipitation products.

Reanalyses have some advantages in analyzing trends. The complexity of describing and understanding trends is multi-faceted, and involves more than simply changes in average quantities over time. Precipitation trends, for example, can be examined in the context of the details of precipitation probability distributions rather than total precipitation amount (Zolina *et al.*, 2004). Observed precipitation trends in the United States reflect more than just an increase in the average itself, being largely related to increases in extreme and heavy rainfall events (Karl and Knight, 1998). Heavier rainfall events seem to be decreasing over tropical land during the last 20 years, a trend that appears to be captured by reanalyses (Takahashi *et al.*, 2006). Warming trends often reflect nighttime warming rather than warming throughout the full 24-hour day (Karl *et al.*, 1991). Precipitation and temperature statistics are fundamentally tied together (Trenberth and Shea, 2005); therefore, their trends should not be studied in isolation.

Given these and other examples of trend complexity, one advantage of a reanalysis dataset becomes clear: a proper analysis of the mechanisms of climate trends requires substantial data, and only a reanalysis provides self-consistent datasets that are complete in space and time over several decades. Given Figures 2.13 to 2.16, future reanalyses need to be improved to support robust trend estimation, particularly for precipitation. However, for many purposes, the comprehensive fields generated by reanalyses, together with their continuity (*i.e.*, no gaps in time, which are a common feature in observational data) and area coverage, provide value for understanding the causes of trends beyond what can be gained from observational datasets alone. For example, by providing trend estimates for midlatitude circulation patterns and other weather elements (features that tend to have a robust signal in reanalyses; see Section 2.4), reanalyses can provide insights into the nature of observed surface temperature and/or precipitation trends.

2.5 STEPS NEEDED TO IMPROVE CLIMATE REANALYSIS

As discussed previously, there are several reasons why the current approaches to assimilating observations for climate reanalysis can lead to false trends and patterns of climate variability. The instruments used to observe the climate may contain systematic errors, and changes in the types of instruments over time may introduce false trends into the observations. Even if the instruments are accurate, the sampling of the instruments across space and time changes over time and thus may improperly introduce shorter time scale or smaller space scale features, or introduce false jumps into the climate record. In addition, the numerical models used to provide a background estimate of the system state contain systematic errors that can project onto the climate analysis. In the case of the ocean, changes in the quality of the surface meteorological forcing will be an additional source of false trends. The following Section address issues of systematic instrument and data sampling errors as well as model and data assimilation errors as a backdrop for recommending improvements in the way future reanalyses are performed. Specific recommendations are given in Chapter 4.

2.5.1 Instrument and Sampling Issues
Prior to the middle of the twentieth century the atmosphere and ocean observing systems consisted mainly of surface observations of variables such as sea level pressure, winds, and surface temperature, although some upper air observations were already being routinely made early in the twentieth century (Brönnimann

et al., 2005). Much of the marine surface data are contained in the International Comprehensive Ocean-Atmosphere Dataset (ICOADS) (Worley *et al.*, 2005) but more still needs to be included. Considerable surface land data also exist, although these are currently scattered throughout several data archives, including those at the National Climatic Data Center and National Center for Atmospheric Research, and many additional surface datasets still need to be digitized. The state of this surface land data should improve as various land data recovery efforts begin (Compo *et al.*, 2006). Attempts to reconstruct climate for the first half of the twentieth century must rely on these surface observations almost exclusively and thus these data recovery efforts are very important (Whitaker *et al.*, 2004; Compo *et al.*, 2006).

In 1936, the U.S. Weather Bureau began operational use of the balloon-deployed radiosonde instrument, providing routine information for atmospheric pressure, temperature, humidity, and wind direction and speed used in daily weather forecasts. By the time of the International Geophysical Year of 1958, the radiosonde network expanded globally to include Antarctica and became recognized as a central component of the historical observation network that climate scientists could use to study climate. As a climate observation network, radiosondes suffer from two major types of problems. First, the instruments contain internal systematic errors (Haimberger, 2007). For example, the widely used Vaisala radiosondes exhibit a tendency toward dryness that needs to be removed (Zipser and Johnson, 1998; Wang *et al.*, 2002). Second, some radiosonde stations

Attempts to reconstruct climate for the first half of the twentieth century must rely on land surface observations almost exclusively and thus data recovery efforts are very important.

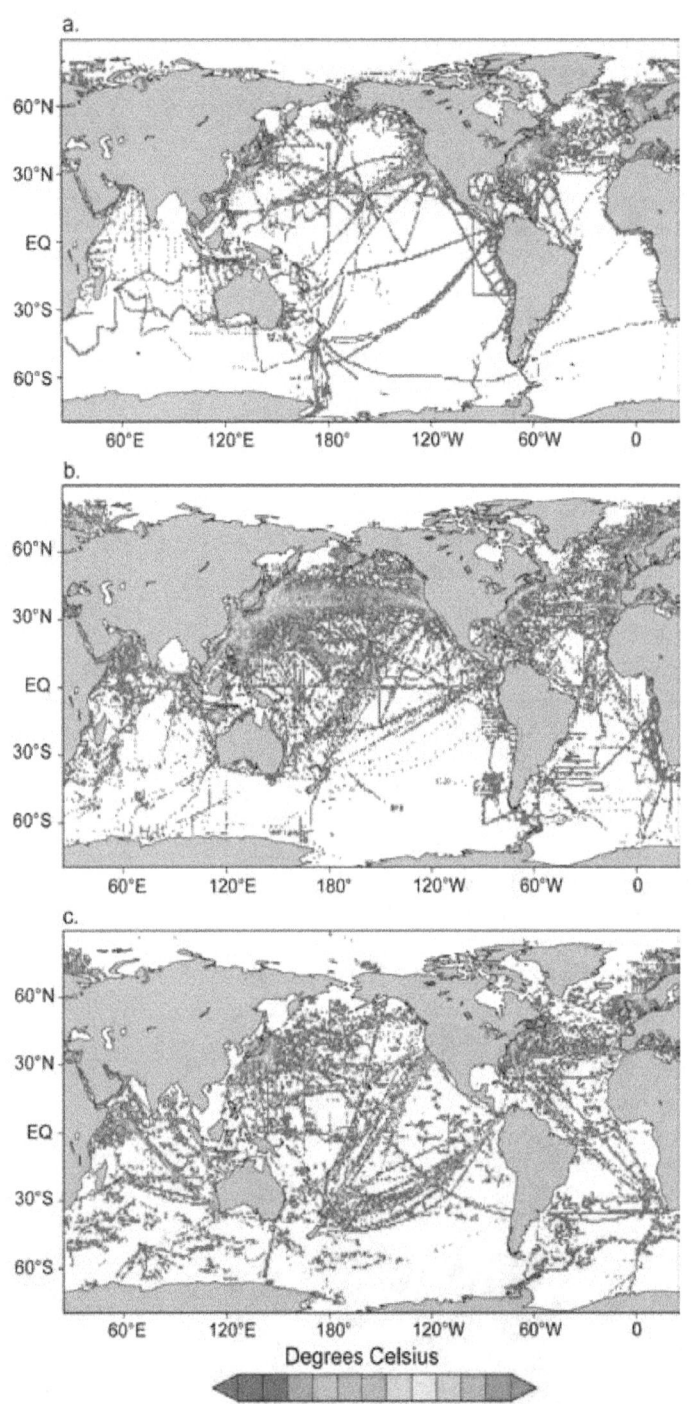

Figure 2.19 Distribution of temperature profile observations in the World Ocean Database extending from the surface of the ocean to 150 meter depth showing 40,000 profiles for 1960 (panel a), 105,000 profiles for 1980 (panel b), and 106,000 profiles for 2004 (panel c) (<http://www.nodc.noaa.gov/OC5/indprod.html>).

have moved to different locations, introducing inconsistencies into the record (Gaffen, 1994).

Two additional observing systems were added to the existing system in the 1970s. Aircraft observations increased in 1973, along with some early satellite-based temperature observations. In 1978, the number of observations increased dramatically in preparation for the First GARP Global Experiment, known as FGGE. The increased observation coverage included three satellite-based vertical temperature sounder instruments (MSU/HIRS/SSU), cloud-tracked winds, and the expansion of aircraft observations and surface observations from ocean drifting buoys. The impact of these additional observations (especially in the Southern Hemisphere) has been noted in the NCEP/NCAR and NCEP/DOE reanalyses (Kalnay *et al.*, 1996; Kistler *et al.*, 2001).

Currently the global radiosonde network consists of about 900 stations, although most radiosondes are launched from continents in the Northern Hemisphere. Of these, there are approximately 600 sonde ascents at 00:00 UTC (Coordinated Universal Time) and 600 ascents at 12:00 UTC, with many from stations that launch the radiosondes only once per day. Most of these launches produce vertical profiles of variables that extend only into the lowest levels of the stratosphere (about six miles above the Earth's surface), at which height the balloons burst. A further troubling aspect of the radiosonde network is the recent closure of stations, especially in Africa, where the network is especially sparse.

As indicated above, the number of atmospheric observations increased dramatically in the 1970s with the introduction of remote sensed temperature retrievals, along with a succession of ancillary measurements (*e.g.*, Figure 2.1). Temperature retrievals are made by observing the intensity of upwelling radiation in the microwave and infrared bands and then using physical models to relate these intensity measurements to a particular temperature profile. The issue of unknown systematic errors in the observations and the need for redundant observations has been highlighted in recent years by a false cooling trend detected in microwave tropospheric temperature retrievals. This false cooling trend has recently been corrected by properly accounting for the effects of orbital decay (Mears *et al.*, 2003).

The ocean observing system has also undergone a gradual expansion of *in situ* observations (*i.e.*, measurements obtained through direct contact with the ocean), followed by a dramatic increase of satellite-based observations (Figures 2.19 and 2.20).

Prior to 1970, the main instrument for measuring subsurface ocean temperature was the mechanical bathythermograph, an instrument primarily deployed along trade shipping routes in the Northern Hemisphere, which recorded temperature only in the upper 280 meters, well above the oceanic thermocline (a thin layer in which temperature changes more rapidly with depth than it does in the layers above or below) at most locations. In the late 1960s the expendable bathythermograph (XBT) was introduced. In addition to being much easier to deploy, the XBT typically records temperature to a depth of 450 meters or 700 meters. Since the late 1980s, moored thermistor arrays have been deployed in the tropical oceans, beginning with the TAO/Triton array of the tropical Pacific, expanding into the Atlantic (PIRATA) in 1997, and most recently into the tropical Indian Ocean. These surface moorings typically measure temperature and, less often, salinity at depths to 500 meters.

Two major problems have been discovered in the historical ocean temperature sampling record. First, much of the data were missing from the oceanographic centers; however, this problem is improving. The 1974 version of the World Ocean Atlas contained 1.5 million profiles. Thanks to great efforts by Global Oceanographic Data Archaeology and Rescue (GODAR) the latest release of the World Ocean Database (WOD2005) contains nearly 8 million profiles (Boyer *et al.*, 2006). Such data archaeology and rescue work needs to be

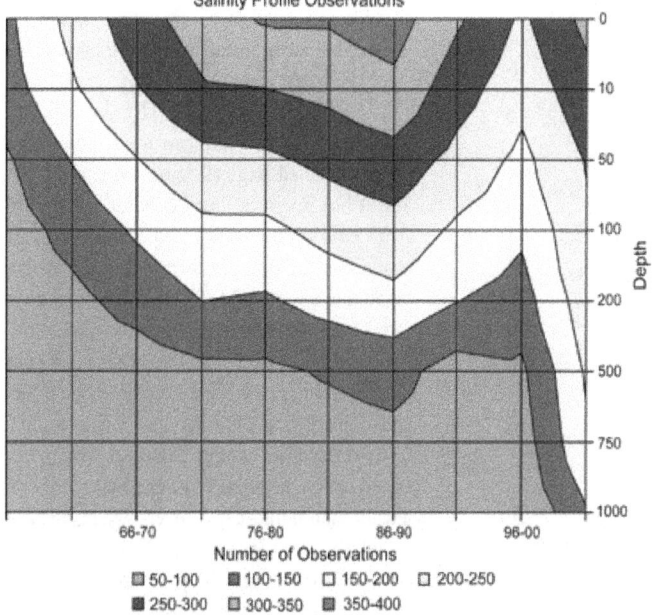

Salinity Profile Observations

Figure 2.20 Distribution of salinity observations as a function of depth and time in the upper 1000 meters from the World Ocean Database 2001 (Carton and Giese, 2008). The decrease in salinity observations in 1974 resulted from the closure of ocean weather stations, while the decrease in the mid 1990s resulted from the end of the World Ocean Circulation Experiment and from the effects of the time delay in transferring salinity observations into the data archives. The recent increase in salinity observations is due to the deployment of the Argo array. Argo is a global array of free-drifting profiling floats that measures the temperature and salinity of the upper 2000 meters of the ocean.

continued. Second, similar to the atmospheric radiosonde, the XBT instrument was not designed for climate monitoring. It is now known that XBT profiles underestimate the depth of the measurement by 1 to 2.5 percent of the actual depth (Hanawa *et al.*, 1995). Unfortunately, the compensating drop-rate correction differs for different varieties of XBTs, and less than half of the XBT observations identify the variety used. Some of the XBT observations collected since the late 1990s have had a drop-rate correction applied without accompanying documentation, while there is evidence that the drop-rate error has changed over time, being higher in the 1970s compared with other time periods (AchutaRao *et al.*, 2007).

For the last half of the twentieth century the main instrument for collecting deep ocean temperature and salinity profiles was the Salinity Temperature Depth or Conductivity Temperature Depth (CTD) sensor. The CTD profiles are accurate, but there are five times fewer CTD profiles compared to the number

Scientists can to al large extent only speculate about the historical changes in deep ocean circulation for the last half of the twentieth century.

of XTB profiles. As a result, scientists can to a large extent only speculate about the historical changes in deep circulation.

Since 2003 a new international observing program called Argo (Roemmich and Owens, 2000) has revolutionized ocean observation. Argo consists of a set of several thousand autonomous drifting platforms that are mainly located at about 1000 meter depth. At regular intervals, generally ten days, the Argo drifters sink and then rise to the surface, recording a profile of temperature and salinity, which is then transmitted via satellite to data archival centers. The introduction of Argo has greatly increased ocean coverage in the Southern Hemisphere as a whole and at mid-depths everywhere, and also greatly increased the number of salinity observations. Argo is gradually being expanded to measure variables such as oxygen levels, which are important for understanding the movement of greenhouse gases .

Satellite remote sensing has further expanded the ocean observing system. This process began in the 1980s with the introduction of infrared and microwave sensing of sea surface temperature, followed by the introduction of continuous radar observations of sea level in the early 1990s, and then by regular surface wind observations from satellite-based scatterometers in the late 1990s. Scatterometers use the radar backscatter from wind-driven ripples on the ocean surface to provide information on wind speed and direction.

The availability of ocean datasets as well as general circulation models of the ocean has led to considerable interest in the development of ocean reanalyses (see Table 2.3). The techniques used are analogous to those used for the atmosphere. One example is the Simple Ocean Data Assimilation (SODA) ocean reanalysis by Carton *et al.* (2000). Like its atmospheric counterpart, this reanalysis shows distinctly different climate variability when satellite data is included.

It is important to address issues regarding the collection and interpretation of reanalysis-relevant land surface data. First, global *in situ* measurements of land states (*e.g.*, soil moisture, snow, ground temperature) are essentially non-existent. Scattered measurements of soil moisture data are available in Asia (Robock *et al.*, 2000), and snow measurement networks provide useful snow information in certain regions (*e.g.*, SNOTEL, <www.wcc.nrcs.usda.gov/snotel/>), but grid-scale *in situ* averages that span the globe are unavailable. Satellite data provide global coverage; however, they have limitations. Even the most advanced satellite-based observations can only measure soil moisture several centimeters into the soil, and not at all under

---

BOX 2.2: Modern Era Retrospective-Analysis for Research and Applications (MERRA)

The NASA/Global Modeling and Assimilation Office (GMAO) atmospheric global reanalysis project is called the Modern Era Retrospective-Analysis for Research and Applications (MERRA). MERRA (Bosilovich *et al.*, 2006) is based on a major new version of the Goddard Earth Observing System Data Assimilation System (GEOS-5), that includes the Earth System Modeling Framework (ESMF)-based GEOS-5 AGCM and the new NCEP unified grid-point statistical interpolation (GSI) analysis scheme developed as a collaborative effort between NCEP and the GMAO.

MERRA supports NASA Earth science by synthesizing the current suite of research satellite observations in a climate data context (covering the period 1979 to present), and by providing the science and applications communities with of a broad range of weather and climate data, with an emphasis on improved estimates of the hydrological cycle.

MERRA products consist of a host of prognostic and diagnostic fields including comprehensive sets of cloud, radiation, hydrological cycle, ozone, and land surface diagnostics. A special collection of data files are designed to facilitate off-line forcing of chemistry/aerosol models. The model or native resolution of MERRA is 0.67° longitude by 0.5° latitude with 72 levels extending to a pressure of 0.01 hectoPascals (hPa). Analysis states and two-dimensional diagnostics will be made available at the native resolution, while many of the three-dimensional diagnostics will be made available on a coarser 1.25° latitude, 1.25° longitude grid. Further information about MERRA and its status may be found at <http://gmao.gsfc.nasa.gov/research/merra/>.

dense vegetation (Entekhabi *et al.*, 2004). Also, existing satellite-based estimates of surface soil moisture, as produced from different sensors and algorithms, are not consistent (Reichle *et al.*, 2007), implying the need for bias correction. Time-dependent gravity measurements may provide soil moisture at deeper levels, but only at spatial scales much coarser than those needed for reanalysis (Rodell *et al.*, 2007). Snow cover data from satellite are readily available, but the estimation of total snow amount from satellite data is subject to significant uncertainty (Foster *et al.*, 2005).

There are now a number of recommendations that have been put forth by the scientific community (*e.g.*, Schubert *et al.*, 2006) in order to make progress on issues regarding data quality and improvement of the world's inventories of atmospheric, ocean, and land observations. These include the need for all major data centers to prepare inventories of observations needed for reanalysis, to form collaborations that can sustain frequent data upgrades and create high quality datasets from all instruments useful for reanalyses, to develop improved record tracking control for observations, and to further improve the use of information about the quality of the reanalyses targeted especially for data providers/developers. Furthermore, the observational, reanalysis, and climate communities should take a coordinated approach to further optimizing the usefulness of reanalysis for climate. These recommendations have now been considered by the WCRP Observations and Assimilation Panel (WOAP) and the Global Climate Observing System (GCOS)/WCRP Atmospheric Observations Panel for Climate.

2.5.2 Modeling and Data Assimilation Issues

False trends may be introduced into the reanalyses by systematic errors in the models used to provide background estimates for data assimilation and by incomplete modeling of those systematic errors in the data assimilation algorithm. Atmospheric models include numerical representations of the primitive equations of motion along with parameterizations of small-scale processes such as radiation, turbulent fluxes, and precipitation. Model integrations begin with some estimate of the initial state, along with boundary values of solar radiation

and sea surface temperature, and are integrated forward in time. While initial global reanalyses (Table 2.1) had resolutions of about 100 to 200 kilometers, the latest reanalysis efforts, NASA's Modern Era Retrospective-Analysis for Research and Applications, MERRA, (see Box 2.2), and NOAA's Reanalysis and Reforecasts of the NCEP Climate Forecast System, CFSRR, (see Box 2.3) have horizontal resolutions of about 50 kilometers or less. Regional models have much finer resolution, currently approaching one kilometer, and time steps of seconds. Improvements in resolution have improved representation of physical processes such as the strength and position of storm tracks and thus have improved simulation of local climate variability and reduced model bias.

Despite these increases in resolution, many important physical processes still cannot be explicitly resolved in current global models, such as convection, cloud formation, and precipitation in the form of both water and ice. Therefore, these processes must be parameterized, or estimated from other, presumably more accurately simulated, model variables. Inaccuracies in these parameterizations are a major source of uncertainty in numerical simulation of the atmosphere and are a cause of false trends, or bias, in atmospheric models. In addition, the presence of atmospheric instabilities (*e.g.*, Farrell, 1989; Palmer, 1988) will lead to model forecast errors.

Ocean models also include representations of primitive equations, with parameterizations for processes such as mixing and sea ice physics. Ocean models exchange thermodynamic, radiative, and momentum fluxes with the atmosphere. Horizontal resolution of current global ocean models is approaching 10 kilometers in order to resolve the complex geometry of the ocean basins and the oceanic mesoscale. Despite this fine resolution, such models still exhibit systematic errors, suggesting that the small horizontal and vertical scales upon which key processes such as vertical mixing, convection, and sea ice formation are still not being resolved (Smith *et al.*, 2000).

In most analyses, the fluxes between ocean and atmosphere are one way because the ocean reanalysis is controlled partly by atmospheric

> Improvements in model resolution have improved representation of physical processes such as the strength and position of storm tracks and thus have improved simulation of local climate variability and reduced model bias.

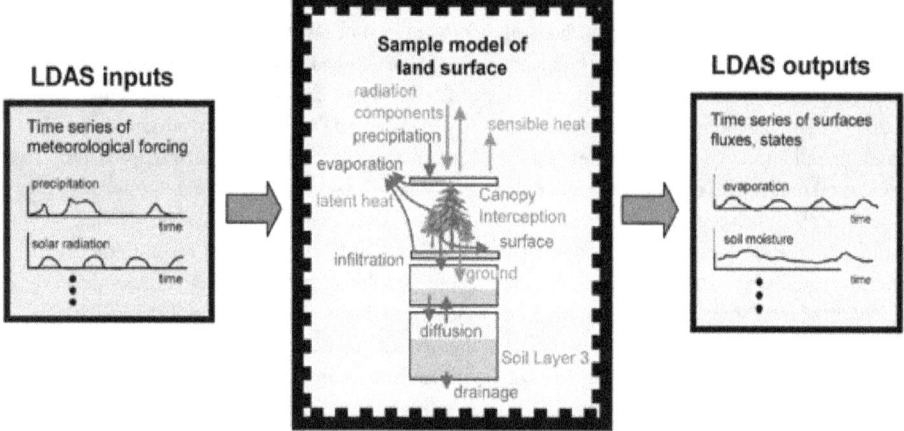

Figure 2.21 Schematic showing the inputs and outputs of a typical Land Data Assimilation System (LDAS) project.

Data assimilation provides a general way to correct a background estimate of the state of the atmosphere, ocean, and land surface that is consistent with available observations.

fluxes, while the atmospheric reanalysis is controlled partly by sea surface temperatures that are specified from observations. Thus, the fluxes in the reanalyses computed for the ocean and for the atmosphere, which should be identical, are in practice substantially different. Carrying out both reanalyses in a fully interconnected atmosphere/ocean model would ensure consistency; however, the surface exchanges are less constrained and thus, initial efforts at a combined analysis have been found to contain considerable systematic errors in both the atmosphere and the ocean (Collins *et al.*, 2006; Delworth *et al.*, 2006). A major challenge in the future will be to correct these systematic errors and subsequently develop consistent and accurate atmosphere/ocean reanalyses. NCEP is currently carrying out the first weakly coupled ocean-atmosphere reanalysis; results are encouraging but it is too early to know the extent to which the fluxes and trends are reliable (Box 2.3).

The land surface component of an atmospheric model also provides fluxes of heat, water, and radiation at the Earth's surface. The major difficulty in producing realistic land fluxes is the large amount of variability (*e.g.*, in topography, vegetation character, soil type, and soil moisture content) across areas (relative to that found in the atmosphere or ocean) in the properties that control these fluxes. These variabilities are difficult to accurately model for two reasons. First, given the area resolutions used for global reanalyses (now and in the foreseeable future), the physical processes that control the

land surface fluxes cannot be properly resolved and therefore the small-scale processes must be parameterized. Second, there are few high resolution global measurements, which are required for many of the relevant land properties.

Despite these limitations, land models have been used in numerous Land Data Assimilation System (LDAS) projects. The current LDAS approach is to drive regional or global arrays of land surface models with observations-based meteorological forcing (*e.g.*, precipitation, radiation) rather than with forcing from an atmospheric model. This allows the land models to evolve their soil moisture and temperature states to presumably realistic values and to produce surface moisture and heat fluxes for diagnostic studies (Figure 2.21).

A list of some current LDAS projects is provided in Table 2.4. The LDAS framework is amenable to true assimilation, in which satellite-derived fields of soil moisture, snow, and temperature are incorporated into the gridded model integrations using new techniques (*e.g.*, Reichle and Koster, 2005; Sun *et al.*, 2004).

Data assimilation provides a general way to correct a background estimate of the state of the atmosphere, ocean, and land surface that is consistent with available observations (Kalnay, 2003; Wunsch, 2006). However, most current data assimilation algorithms make several assumptions either for efficiency or from lack of information, limiting their effectiveness. These assumptions include: (1) that any systematic

Table 2.4 A partial list of current Land Data Assimilation System (LDAS) projects.

Project	Sponsor(s)	Spatial Domain	Unique Aspects	Reference	Project website
GSWP-2	GEWEX	Global, 1°	Separate datasets produced by at least 15 land models for the period 1986 to 1995	Dirmeyer *et al.* (2006)	<http://www.iges.org/gswp2/>
GLDAS	NASA, NOAA	Global, .25° to ~2°	Multiple land models; near-real-time data generation	Rodell *et al.* (2004)	<http://ldas.gsfc.nasa.gov/>
NLDAS	Multiple Institutions	Continental U.S., 0.125°	Multiple land models; near-real-time data generation	Mitchell *et al.* (2004)	<http://ldas.gsfc.nasa.gov/>
ELDAS and ECMWF follow-on	European Commission	Europe, 0.2°	True data assimilation of air temperature and humidity in some versions	Van den Hurk (2002); Van den Hurk *et al.* (2008)	<http://www.knmi.nl/samenw/eldas/>

trends or biases in the observation measurements or sampling have been identified and corrected; (2) that the forecast model is unbiased; and (3) that the error statistics, such as the model forecast error, have linear, Gaussian (normally distributed) characteristics.

Several changes can be made to improve these assumptions. Systematic errors introduced by expansions of the observing system can be reduced by repeating the reanalysis with a reduced, but more consistent dataset, excluding,

for example, satellite observations. An extreme version of this approach is to use only surface observations (Compo *et al.*, 2006). In this case, atmospheric reanalysis methods would need to make better use of historical surface observations from land stations and marine platforms. These records include existing climate datasets, such as daily or monthly air temperature, pressure, humidity, precipitation, and cloudiness, which have already undergone extensive quality control for the purpose of climate variability and trend applications.

BOX 2.3: Climate Forecast System Reanalysis and Reforecast Project (CFSRR)

The New Reanalysis and Reforecasts of the NCEP Climate Forecast System (CFSRR) is a major upgrade to the coupled atmosphere/ocean/land Climate Forecast System (CFS; Saha *et al.*, 2006). This upgrade is planned for January 2010 and involves changes to all components of the CFS, including the NCEP atmospheric Gridded Statistical Interpolation scheme (GSI), the NCEP atmospheric Global Forecast System (GFS), the NCEP Global Ocean Data Assimilation System (GODAS), which includes the use of the new GFDL MOM4 Ocean Model, and the NCEP Global Land Data Assimilation System (GLDAS), which includes the use of a new NCEP NOAA Land model.

There are two essential components to this upgrade: a new reanalysis of atmosphere, ocean, land, and sea ice, and a complete reforecast of the new CFS. The new reanalysis will be conducted for the 31-year period (1979 to 2009). The reanalysis system includes an atmosphere with high horizontal (spectral T382, about 38 km) and vertical (64 sigma-pressure hybrid levels) resolution, an ocean with 40 levels in the vertical to a depth of 4737 meters and a horizontal resolution of 0.25° at the tropics, tapering to a global resolution of 0.5° northwards and southwards of 10°N and 10°S, respectively, an interactive sea ice model, and an interactive land model with four soil levels.

In addition to the higher horizontal and vertical resolution of the atmosphere, the key differences from the previous NCEP global reanalysis are that the guess forecast will be generated from an interconnected atmosphere-ocean-land-sea ice system, and that radiance measurements from the historical satellites will be assimilated.

As scientists continue to improve coupled models, joint assimilation between atmosphere, ocean, and land components should ensure greater consistency of model states across the components because the states of the systems would be allowed to evolve together.

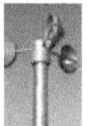

Systematic errors in the models may be explicitly accounted for and thus potentially corrected in the data assimilation algorithm (*e.g.*, Dee and da Silva, 1998; Danforth *et al.*, 2007). However, additional work is needed to improve bias modeling. In addition to estimating and reducing bias, there is a need to improve the representation of error covariances, and to provide improved estimates of the uncertainties in all reanalysis products. New techniques (*e.g.*, the Ensemble Kalman Filter) are being developed that are both economical and able to provide such estimates (*e.g.*, Tippett *et al.*, 2003; Ott *et al.*, 2004).

Looking ahead, a promising pathway for improved reanalyses is the development of coupled data assimilation systems, along with methods to correct for the tendency of coupled models to develop bias. In this case, the observed atmosphere, ocean, and land states are assimilated jointly into the atmosphere, ocean, and land components of a fully coupled climate system model; however, the substantial bias in current coupled models makes this a significant challenge. Nevertheless, as scientists continue to improve coupled models, this joint assimilation should ensure greater consistency of model states across the components because the states would be allowed to evolve together. For example, a satellite-based correction to a soil moisture value would be able to impact and thereby potentially improve overlying atmospheric moisture and temperature states. The overall result of coupled assimilation would presumably be a more reliable and more useful reanalysis product. Several efforts are moving toward coupled data assimilation in the United States. These are focused primarily on developing more balanced initial conditions for the seasonal and longer forecast problem, and include the Climate Forecast System Reanalysis and Reforecast (CFSRR, see Box 2.3) project at NCEP and an ensemble-based approach being developed at NOAA's Geophysical Fluid Dynamics Laboratory (GFDL) (Zhang *et al.*, 2007). Also, the GMAO is utilizing both the MERRA product (Box 2.2) and an ocean data assimilation system to explore data assimilation in a fully coupled climate model.

CHAPTER 3

Attribution of the Causes of Climate Variations and Trends over North America during the Modern Reanalysis Period

Convening Lead Author: Martin Hoerling, NOAA/ESRL

Lead Authors: Gabriele Hegerl, Edinburgh Univ.; David Karoly, Univ. of Melbourne; Arun Kumar, NOAA; David Rind, NASA

Contributing Author: Randall Dole, NOAA/ESRL

KEY FINDINGS

- Significant advances have occurred over the past decade in capabilities to attribute causes for observed climate variations and change.
- Methods now exist for establishing attribution for the causes of North American climate variations and trends due to internal climate variations and/or changes in external climate forcing.

Annual, area-averaged change since 1951 across North America shows:
- Seven of the warmest ten years for annual surface temperatures since 1951 have occurred in the last decade (1997 to 2006).
- The 56-year linear trend (1951 to 2006) of annual surface temperature is +0.90°C ±0.1°C (1.6°F ± 0.2°F).
- Virtually all of the warming since 1951 has occurred after 1970.
- More than half of the warming is *likely* the result of anthropogenic greenhouse gas forcing of climate change.
- Changes in ocean temperatures *likely* explain a substantial fraction of the anthropogenic warming of North America.
- There is no discernible trend in average precipitation since 1951, in contrast to trends observed in extreme precipitation events (CCSP, 2008).

Spatial variations in annually-averaged change for the period 1951 to 2006 across North America show:
- Observed surface temperature change has been largest over northern and western North America, with up to +2°C (3.6°F) warming in 56 years over Alaska, the Yukon Territories, Alberta, and Saskatchewan.
- Observed surface temperature change has been smallest over the southern United States and eastern Canada, where no significant trends have occurred.
- There is *very high* confidence that changes in atmospheric wind patterns have occurred, based upon reanalysis data, and that these wind pattern changes are the *likely* physical basis for much of the spatial variations in surface temperature change over North America, especially during winter.
- The spatial variations in surface temperature change over North America are *unlikely* to be the result of anthropogenic forcing alone.
- The spatial variations in surface temperature change over North America are *very likely* influenced by variations in global sea surface temperatures through the effects of the latter on atmospheric circulation, especially during winter.

Spatial variations of seasonal average change for the period 1951 to 2006 across the United States show:

- Six of the warmest 10 summers and winters for the contiguous United States average surface temperatures from 1951 to 2006 have occurred in the last decade (1997 to 2006).
- During summer, surface temperatures have warmed most over western states, with insignificant change between the Rocky Mountains and the Appalachian Mountains. During winter, surface temperatures have warmed most over northern and western states, with insignificant change over the central Gulf of Mexico and Maine.
- The spatial variations in summertime surface temperature change are *unlikely* to be the result of anthropogenic greenhouse forcings alone.
- The spatial variations and seasonal differences in precipitation change are *unlikely* to be the result of anthropogenic greenhouse forcings alone.
- Some of the spatial variations and seasonal differences in precipitation change and variations are *likely* the result of regional variations in sea surface temperatures.

An assessment to identify and attribute the causes of abrupt climate change over North America for the period 1951 to 2006 shows:

- There are limitations for detecting rapid climate shifts and distinguishing these shifts from quasi-cyclical variations because current reanalysis data only extends back to the mid-twentieth century. Reanalysis over a longer time period is needed to distinguish between these possibilities with scientific confidence.

An assessment to determine trends and attribute causes for droughts for the period 1951 to 2006 shows:

- It is *unlikely* that a systematic change has occurred in either the frequency or area coverage of severe drought over the contiguous United States from the mid-twentieth century to the present.
- It is *very likely* that short-term (monthly-to-seasonal) severe droughts that have impacted North America during the past half-century are mostly due to atmospheric variability, in some cases amplified by local soil moisture conditions.
- It is *likely* that sea surface temperature anomalies have been important in forcing long-term (multi-year) severe droughts that have impacted North America during the past half-century.
- It is *likely* that anthropogenic warming has increased drought impacts over North America in recent decades through increased water stresses associated with warmer conditions, but the magnitude of the effect is uncertain.

INTRODUCTION

Increasingly, climate scientists are being asked to go beyond descriptions of *what* the current climate conditions are and how they compare with the past, to also explain *why* climate is evolving as observed; that is, to provide attribution of the causes for observed climate variations and change.

Today, a fundamental concern for policy makers is to understand the extent to which anthropogenic factors and natural climate variations are responsible for the observed evolution of climate. A central focus for such efforts, as articulated in the Intergovernmental Panel on Climate Change (IPCC) Assessment Reports (IPCC, 2007a) has been to establish the cause, or causes, for globally averaged temperature increases over roughly the past century. However, requests for climate attribution far transcend

> Today, a fundamental concern for policy makers is to understand the extent to which anthropogenic factors and natural climate variations are responsible for the observed evolution of climate.

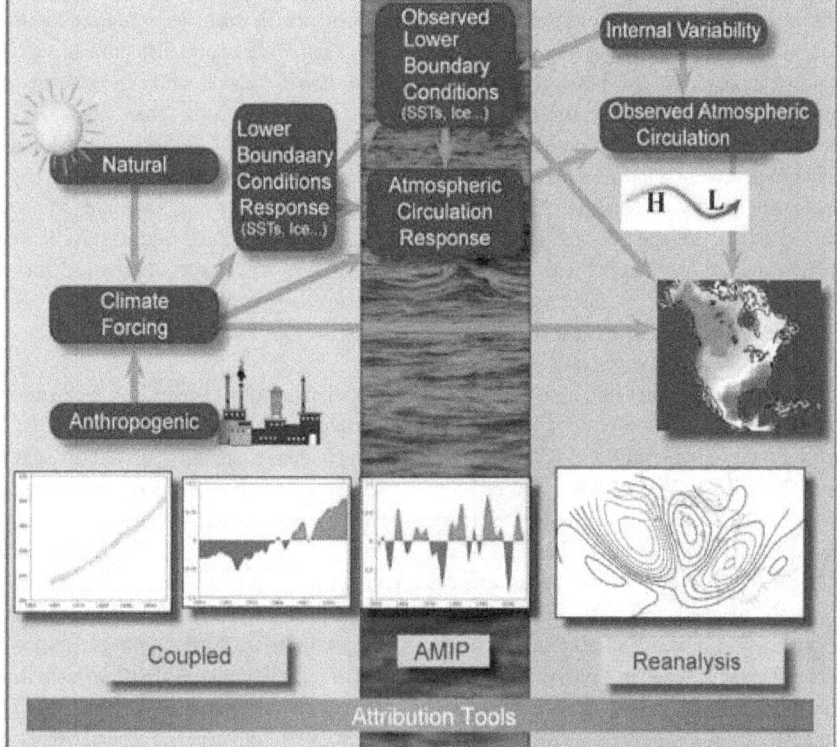

Figure 3.1 Schematic illustration of the datasets and modeling strategies for performing attribution. The map of North America on the right side displays a climate condition whose origin is in question. Various candidate causal mechanisms are illustrated in the right-to-left sequences of figures, together with the attribution tool. Listed above each in maroon boxes is a plausible cause that could be assigned to the demonstrated mechanism depending upon the diagnosis of forcing-response relationships derived from attribution methods. The efficacy of the first mechanism is tested, often empirically, by determining consistency with patterns of atmospheric variability, such as the teleconnection processes (climate anomalies over different geographical regions that are linked by a common cause) identifiable from reanalysis data. This step places the current condition within a global and historical context. The efficacy of the second mechanism tests the role of boundary forcings, most often with atmospheric models (e.g., Atmospheric Model Intercomparison Project, AMIP). The efficacy of the third mechanism tests the role of natural or anthropogenic influences, most often with linked ocean-atmosphere models. The processes responsible for the climate condition in question may, or may not, involve teleconnections, but may result from local changes in direct radiative effect on climate change or other near-surface forcing such as from land surface anomalies. The lower panels illustrate the representative processes: from left-to-right; time-evolving atmospheric carbon dioxide at Mauna Loa, Hawaii, the warming trend over several decades in tropical Indian Ocean/West Pacific warm pool sea surface temperatures (SSTs), the yearly SST variability over the tropical east Pacific due to the El Niño-Southern Oscillation (ENSO), the atmospheric pattern over the North Pacific/North America referred to as the Pacific North American (PNA) teleconnection.

global temperature change alone, with notable interest in explaining regional temperature variations and the causes for high-impact climate events, such as the recent multi-year drought in the western United States and the record setting U.S. warmth in 2006. For many decision makers who must assess potential impacts and management options, a particularly important question is: What are the causes for regional and seasonal differences in climate variations and trends, and how well do we understand them? For example, is the recent drought in the western United States due mainly to factors internal to the climate system (*e.g.*, the sea surface temperature variations associated with ENSO), in which case a return toward previous climate conditions might be anticipated, or is it a manifestation of a longer-term trend toward increasing aridity in the region that is driven primarily by anthropogenic forcing? Why do some droughts last longer than others? Such examples illustrate that, in order to support informed decision making, the capability to attribute causes for past and current climate conditions can be a major consideration.

The recently completed IPCC Fourth Assessment Report (AR4) from Working Group I contains a full chapter (Chapter 9) devoted to the topic "Understanding and Attributing Climate Change" (IPCC, 2007a). This Chapter attempts to minimize overlap with the IPCC Report by focusing on a subset of questions of particular interest to the U.S. public, decision makers, and policy makers that may not have been covered in detail (or in some cases, at all) in the IPCC Report. The specific emphasis here is on present scientific capabilities to attribute the causes for observed climate variations and change over North America. For a more detailed discussion of attribution, especially for other regions and at the global scale, the interested reader is referred to Chapter 9 of the AR4 Working Group I Report (IPCC, 2007a).

Figure 3.1 illustrates methods and tools used in climate attribution. The North American map (right side) shows an observed surface condition, the causes of which are sought. A roadmap for attribution involves the systematic probing of cause-effect relationships. Plausible factors that contribute to the change are identified along the top of Figure 3.1 (maroon boxes),

and arrows illustrate connections among these as well as pathways for explaining the observed condition.

The attribution process begins by examining conditions of atmospheric wind patterns (also called circulation patterns) that coincide with the North American surface climate anomaly. It is possible, for instance, that the surface condition evolved concurrently with a change in the tropospheric jet stream, such as accompanies the Pacific-North American pattern (see Chapter 2). Reanalysis data are essential for this purpose because they provide a global description of the state of the troposphere (the lowest region of the atmosphere which extends from the Earth's surface to around 10 kilometers, or about 6 miles, in altitude) that is physically consistent in space and time. Although reanalysis can illuminate a connection between atmospheric circulation patterns and surface climate, it may not directly implicate the causes, that is, provide attribution.

Additional tools are often needed to explain the atmospheric circulation pattern itself. Is it, for instance, due to chaotic internal atmospheric variations, or is it related to forcing external to the atmosphere (*e.g.*, changes in sea surface temperatures or solar forcing)? The middle column in Figure 3.1 illustrates the common approach used to assess the forcing-response associated with Earth's surface boundary conditions (physical conditions at a given boundary), in particular sea surface temperatures. The principal tool is atmospheric general circulation models that are forced, that is, are subjected to a specific influence (see Box 3.2), for example, a specified history of sea surface temperatures as boundary conditions (Gates, 1992). Reanalysis would continue to be important in this stage of attribution in order to evaluate the suitability of the models as an attribution tool, including the realism of simulated circulation variability (Box 3.1).

In the event that diagnosis of the Atmospheric Model Intercomparison Project (AMIP) simulation fails to confirm a role for Earth's lower boundary conditions, then two plausible explanations for the atmospheric circulation (and its associated North American surface condition) remain. One explanation is that it was due to

For many decision makers who must assess potential impacts and management options, a particularly important question is: What are the causes for regional and seasonal differences in climate variations and trends, and how well do we understand them?

chaotic atmospheric variability rather than natural or anthropogenic influences. Reanalysis data would be useful to determine whether the circulation state was within the scope of known variations during the reanalysis record. The other possible explanation is that external natural (*e.g.*, volcanic and solar) or external anthropogenic perturbations may directly have caused the responsible circulation pattern. Coupled ocean-atmosphere climate models would be used to explore the forcing-response relationships involving such external forcings. As illustrated by the left column, coupled models have been widely employed in the Reports of the IPCC. Here again, reanalysis is important for assessing the suitability of this attribution tool, including the realism of simulated ocean-atmosphere variations such as the El Niño-Southern Oscillation (ENSO) and accompanying atmospheric teleconnections (climate anomalies over different geographical regions that are linked by a common cause) that influence North American surface climate (Box 3.1).

If diagnosis of the AMIP simulations confirms a role for Earth's lower boundary conditions, it becomes important to explain the cause for the boundary condition itself. Comparison of the observed sea surface temperatures with coupled model simulations would be the principal approach. If externally-forced models that consider human influences on climate change fail to yield the observed boundary conditions, then the boundary condition may be attributed to chaotic intrinsic coupled ocean-atmosphere variations. If coupled models instead replicate the observed boundary conditions, this establishes a consistency with external forcing as an ultimate cause. (In addition, it is necessary to confirm that the coupled models also generate the atmospheric circulation patterns; that is, to demonstrate that the model result is obtained for the correct physical reason.)

Figure 3.1 illustrates basic approaches applied in the following sections of Chapter 3. It is evident that a physically-based scientific interpretation for the causes of a climate condition requires accurately measured and analyzed features of the time and space characteristics of atmospheric circulation and surface conditions. In addition, the interpretation relies heavily upon the use of

climate models to test candidate cause-effect relations. Reanalysis is essential for both components of such attribution science.

While this Chapter considers the approximate period covered by modern reanalyses (roughly 1950 to the present), datasets other than reanalyses, such as gridded surface station analyses of temperature and precipitation, are also used. The surface conditions illustrated in Figure 3.1 are generally derived from such datasets, and these are extensively used to describe various key features of the recent North American climate variability in Chapter 3. These, together with modern reanalysis data, provide a necessary historical context against which the uniqueness of current climate conditions both at Earth's surface and in the free atmosphere can be assessed.

3.1 CLIMATE ATTRIBUTION AND SCIENTIFIC METHODS USED FOR ESTABLISHING ATTRIBUTION

3.1.1 What is Attribution?

Climate attribution is a scientific process for establishing the principal causes or physical explanation for observed climate conditions and phenomena. Within its Reports, the IPCC states that "attribution of causes of *climate change* is the process of establishing the most likely causes for the detected change with some level of confidence" (IPCC 2007). As noted in the Introduction, the definition is expanded in this Product to include attribution of the causes of observed *climate variations* that may not be unusual in a statistical sense but for which great public interest exists because they produce major societal impacts.

It is useful to outline some general classes of mechanisms that may produce climate variations or change. One important class is *external forcing*, which contains both *natural* and *anthropogenic* sources. Examples of natural external forcing include solar variability and volcanic eruptions. Examples of anthropogenic forcing are changing concentrations of greenhouse gases and aerosols and land cover changes produced by human activities. A second class involves *internal mechanisms* within the climate system that can produce climate

Climate attribution is a scientific process for establishing the principal causes or physical explanation for observed climate conditions and phenomena.

BOX 3.1: Assessing Model Suitability

A principal tool for attributing the causes of climate variations and change involves climate models. For instance, atmospheric models using specified sea surface temperatures are widely used to assess the impact of El Niño on seasonal climate variations. Coupled ocean-atmosphere models using specified atmospheric chemical constituents are widely used to assess the impact of greenhouse gases on detected changes in climate conditions. One prerequisite for the use of models as tools is their capacity to simulate the known leading patterns of atmospheric (and for the coupled models, oceanic) modes of variations. Realism of the models enhances confidence in their use for probing forcing-response relationships, and it is for this reason that an entire chapter of the Intergovernmental Panel on Climate Change (IPCC) Fourth Assessment Report (AR4) is devoted to evaluation of the models for simulating known features of large-scale climate variability. That report emphasizes the considerable scrutiny and evaluations under which these models are being placed, making it "less likely that significant model errors are being overlooked". Reanalysis data of global climate variability of the past half-century provide valuable benchmarks against which key features of model simulations can be meaningfully assessed.

The box figure illustrates a simple use of reanalysis for validation of models that are employed for attribution elsewhere in this report. Chapter 8 of the Working Group I report of IPCC AR4 and the references therein provide numerous additional examples of validation studies of the IPCC coupled models that are used in this SAP. Shown are the leading winter patterns of atmospheric variability, discussed previously in Chapter 2 (Figures 2.8 and 2.9), that have strong influence on North American climate. These are the Pacific-North American pattern (left), the North Atlantic Oscillation pattern (middle), and the El Niño-Southern Oscillation pattern (right). The spatial expressions of these patterns is depicted using correlations between observed (simulated) indices of the PNA, NAO, and ENSO with wintertime 500 hectoPascals geopotential heights derived from reanalysis (simulation) data for 1951 to 2006. Both atmospheric (middle) and coupled ocean-atmospheric (bottom) models realistically simulate the phase and spatial scales of the observed (top) patterns over the Pacific-North American domain. The correlations within the PNA and NAO centers of action are close to those observed indicating the fidelity of the models in generating these atmospheric teleconnections. The ENSO correlations are appreciably weaker in the models than in reanalysis. This is in part due to averaging over multiple models and multiple realizations of the same model. It perhaps also indicates that the tropical-extratropical interactions in these models is weaker than observed, and for the CMIP runs it may also indicate weaker ENSO sea surface temperature variability. These circulation patterns are less pronounced during summer, at which time climate variations become more dependant upon local processes (e.g., convection and land-surface interaction) which poses a greater challenge to climate models.

More advanced applications of reanalysis data to evaluate models include budget diagnoses that test the realism of physical processes associated with climate variations, frequency analysis of the time scales of variations, and multivariate analysis to assess the realism of coupling between surface and atmospheric fields. It should be noted that despite the exhaustive evaluations that can be conducted, model assessments are not always conclusive about their suitability as an attribution tool. First, the tolerance to biases in models needed to produce reliable assessment of cause-effect relationships is not well understood. It is partly for this reason that large multi-model ensemble methods are employed for attribution studies in order to reduce the random component of biases that exist across individual models. Second, even when known features of the climate system are judged to be realistically simulated in models, there is no assurance that the modeled response to increased greenhouse gas emissions will likewise be realistic under future scenarios. Therefore attribution studies (IPCC, Chapter 9) compare observed with climate model simulated change because such sensitivity is difficult to evaluate from historical observations.

variations manifesting themselves over seasons, decades, and longer. Internal mechanisms include processes that are due primarily to interactions within the atmosphere as well as those that involve coupling the atmosphere with various components of the climate system. Climate variability due to purely internal mechanisms is often called *internal variability*.

For attribution to be established, the relationship between the observed climate state and the proposed causal mechanism needs to be demonstrated, and alternative explanations need to be determined as unlikely. In the case of attributing the cause of a climate condition to internal variations, for example, due to ENSO-related tropical east Pacific sea surface conditions, the influence of alternative modes of internal climate variability must also be assessed. Before attributing a climate condition to anthropogenic forcing, it is important to determine whether the climate condition was

BOX 3.1: Assessing Model Suitability *Cont'd*

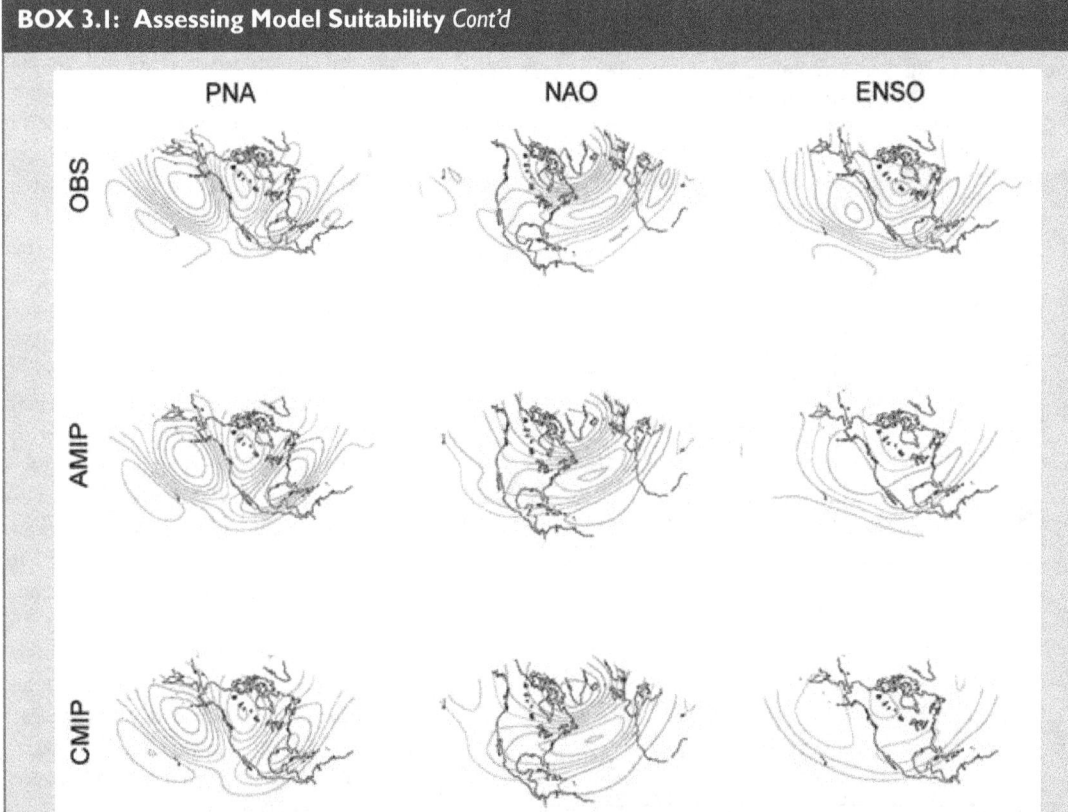

Figure Box 3.1 Temporal correlation between winter season (December, January, February) 500 hectoPascals (hPa) geopotential heights and indices of the leading patterns of Northern Hemisphere climate variability: Pacific-North American (PNA, left), North Atlantic Oscillation (middle), and El Niño-Southern Oscillation (ENSO, right) circulation patterns. The ENSO index is based on equatorial Pacific sea surface temperatures averaged 170°W to 120°W, 5°N to 5°S, and the PNA and NAO indices based on averaging heights within centers of maximum observed height variability following Wallace and Gutzler (1981). Assessment period is 1951 to 2006: observations based on reanalysis data (top), simulations based on atmospheric climate models forced by observed specified sea surface temperature variability (middle), and coupled ocean-atmosphere models forced by observed greenhouse gas, atmospheric aerosols, solar and volcanic variability (bottom). AMIP comprised of 2 models and 33 total simulations. CMIP comprised of 19 models and 41 total simulations. Positive (negative) correlations in red (blue) contours.

unlikely to have resulted from natural external forcing or internal variations alone.

Attribution is associated with the process of explaining the cause of a detected change. In particular, attribution of anthropogenic climate change—the focus of the IPCC Reports (Houghton *et al.*, 1996; IPCC, 2001; IPCC, 2007a)—has the specific objective of explaining a detected climate change that is significantly different from that which could be expected from natural external forcing or internal variations of the climate system. According to the IPCC Third Assessment Report, the attribution requirements for a detected change are: (1) a demonstrated consistency with a combination of anthropogenic and natural external forcings,

and (2) an inconsistency with "alternative, physically plausible explanations of recent climate change that exclude important elements of the given combination of forcings" (IPCC, 2001).

3.1.2 How is Attribution Performed?
The methods used for attributing the causes for observed climate conditions depend on the specific problem or context. To establish the cause, it is necessary to identify possible forcings, determine the responses produced by such forcings, and determine the agreement between the forced response and the observed condition. It is also necessary to demonstrate that the observed climate condition is unlikely to have originated from other forcing mechanisms.

Table 3.1 Acronyms of climate models referenced in this Chapter. All 19 models performed simulations of twentieth century climate change ("20CEN") as well as the 720 parts per million (ppm) stabilization scenario (SRESA1B) in support of the IPCC Fourth Assessment Report (IPCC, 2007a). The ensemble size (ES) is the number of independent realizations of the 20CEN experiment that were analyzed here.

	Model Acronym	Country	Institution	ES
1	CCCma-CGCM3.1(T47)	Canada	Canadian Centre for Climate Modelling and Analysis	1
2	CCSM3	United States	National Center for Atmospheric Research	6
3	CNRM-CM3	France	Météo-France/Centre National de Recherches Météorologiques	1
4	CSIRO-Mk3.0	Australia	CSIRO[a] Marine and Atmospheric Research	1
5	ECHAM5/MPI-OM	Germany	Max-Planck Institute for Meteorology	3
6	FGOALS-g1.0	China	Institute for Atmospheric Physics	1
7	GFDL-CM2.0	United States	Geophysical Fluid Dynamics Laboratory	1
8	GFDL-CM2.1	United States	Geophysical Fluid Dynamics Laboratory	1
9	GISS-AOM	United States	Goddard Institute for Space Studies	2
10	GISS-EH	United States	Goddard Institute for Space Studies	3
11	GISS-ER	United States	Goddard Institute for Space Studies	2
12	INM-CM3.0	Russia	Institute for Numerical Mathematics	1
13	IPSL-CM4	France	Institute Pierre Simon Laplace	1
14	MIROC3.2(medres)	Japan	Center for Climate System Research/NIES[b]/JAMSTEC[c]	3
15	MIROC3.2(hires)	Japan	Center for Climate System Research/NIES[b]/JAMSTEC[c]	1
16	MRI-CGCM2.3.2	Japan	Meteorological Research Institute	5
17	PCM	United States	National Center for Atmospheric Research	4
18	UKMO-HadCM3	United Kingdom	Hadley Centre for Climate Prediction and Research	1
19	UKMO-HadGEM1	United Kingdom	Hadley Centre for Climate Prediction and Research	1

[a] CSIRO is the Commonwealth Scientific and Industrial Research Organization.
[b] NIES is the National Institute for Environmental Studies.
[c] JAMSTEC is the Frontier Research Center for Global Change in Japan.

The methods for signal identification, as discussed in more detail below, involve both empirical analysis of past climate relationships and experiments with climate models in which forcing-response relations are evaluated. Similarly, estimates of internal variability can be derived from the instrumental records of historical data including reanalyses and from simulations performed by climate models in the absence of the candidate forcings. Both empirical and modeling approaches have limitations. Empirical approaches are hampered by the relatively short duration of the climate record, the confounding of influences from various forcing mechanisms, and possible non-physical inconsistencies in the climate record that can result from changing monitoring techniques and analysis procedures (see Chapter 2 for examples of non-physical trends in precipitation due to shifts in reanalysis methods). The climate models are hampered by uncertainties in the representation of physical processes and by coarse spatial resolution, meaning that each grid cell in a global climate model generally covers an area of several hundred kilometers, which can lead to model biases.

In each case, the identified signal (forcing-response relationship) must be robust to these uncertainties, and requires demonstrating that an empirical analysis is both physically meaningful, is insensitive to sample size, and is reproducible when using different climate models. Best attribution practices employ combinations of empirical and numerical approaches using multiple climate models to minimize the effects of possible biases resulting from a single line of approach. Following this approach, Table 3.1 and Table 3.2 lists the observational and model datasets used to generate analyses in Chapter 3.

The specific attribution method can also differ according to the forcing-response relation being probed. As discussed below, three methods have been widely employed. These methods consider different hierarchical links in causal relationships as illustrated in the Figure 3.1 schematic and discussed in Section 3.1.2.1: (1) climate conditions arising from mechanisms internal to the atmosphere; (2) climate conditions forced from changes in atmospheric lower boundary conditions (for example, changes in ocean or land surface conditions); and (3) climate conditions forced externally, whether natural or anthropogenic. In some cases, more than one of these links, or pathways, can be involved. For example, changes in greenhouse gas forcing may induce changes in the ocean component of the climate system. These changing ocean conditions can then force a response in the atmosphere that leads to regional temperature or precipitation changes.

3.1.2.1 SIGNAL DETERMINATION

1) Attribution to internal atmospheric variations

Pioneering empirical research, based only on surface information, discovered statistical linkages between anomalous climate conditions that were separated by continents and oceans (Walker and Bliss, 1932), structures that are referred to today as teleconnection patterns. The North Atlantic Oscillation (NAO), which is a see-saw in anomalous pressure between the subtropical North Atlantic and the Arctic, and the Pacific-North American (PNA) pattern, which is a wave pattern of anomalous climate conditions arching across the North Pacific and North American regions, are particularly relevant to understanding North American climate variations. Chapter 2 illustrates the use of reanalysis data to diagnose the tropospheric wintertime atmospheric circulations associated with a specific phase of the PNA and NAO patterns, respectively. They each have widespread impacts on North American climate conditions as shown by station-based analyses of surface temperature and precipitation anomalies, and the reanalysis data of free atmospheric conditions provides the foundation for a physical explanation of the origins of those fingerprints (physical patterns), (see Section 3.1.2.2). The reanalysis data are also used to validate the realism of atmospheric circulation in climate models, as illustrated in Box. 3.1.

Observations of atmospheric circulation patterns in the free atmosphere fueled theories of the dynamics of these teleconnections, clarifying the origins for their regional surface impacts (Rossby, 1939). The relevant atmospheric circulations represent fluctuations in the semi-permanent positions of high and low pressure centers, their displacements being induced by a variety of mechanisms including

Table 3.2 Datasets utilized in the Product. The versions of these data used in this Product include data through December 2006. The web sites listed below provide URLs to the latest versions of these datasets, which may incorporate changes made after December 2006.

URL Link Information for Data Sets
CRU   HadCRUT3v   Climatic Research Unit of the University of East Anglia and the Hadley Centre of the UK Met Office
\<http://www.cru.uea.uk/cru/data/temperature/\>
NOAA   Land/Sea Merged Temperature   NOAA's National Climatic Data Center (NCDC)
\<http://www.ncdc.noaa.gov/oa/climate/research/anomalies/\>
NASA   Land+Ocean Temperature   NASA's Goddard Institute for Space Studies (GISS)
\<http://data.giss.noaa.gov/gistemp/\>
NCDC   Gridded Land Temperature NOAA's National Climatic Data Center (NCDC)   Gridded Land Precipitation
\<http://www.ncdc.noaa.gov/oa/climate/research/ghcn/\>
NCDCdiv   Contiguous U.S. Climate Division Data (temperature and precipitation)
\<http://www.ncdc.noaa.gov/oa/climate/onlineprod/\>
PRISM   Spatial Climate Gridded Data Sets (temperature and precipitation) Oregon State University's Oregon Climate Service (OCS)
\<http://prism.oregonstate.edu/\>
CHEN   Global Land Precipitation   NOAA's Climate Prediction Center (CPC)
\<http://www.cpc.noaa.gov/products/precip/\>
GPCC   Global Gridded Precipitation Analysis Global Precipitation Climatology Centre (GPCC)
\<http://www/dwd/de/en/FundE/Klima/KLIS/int/GPCC/\>
CMIP3   CMIP3   World Climate Programme's (WCRP's) Coupled Model Intercomparison Project phase 3 (CMIP3) multi-model dataset
\<http://www-pcmdi.llnl.gov/ipcc/\>
Reanalysis   NCEP50   National Centers for Environmental Prediction (NCEP), NOAA, and the National Center for Atmospheric Research (NCAR)
\<http://dss.ucar.edu/pub/reanalysis/data_usr.html/\>
ECHAM4.5   ECHAM4.5
\<http://iridl.ldeo.columbia.edu/SOURCES/.IRI/.FD/.ECHAM4p5/.History/.MONTHLY\>
NASA/NSIPP Runs

anomalous atmospheric heating (*e.g.*, due to changes in tropical rainfall patterns), changes in wind flow over mountains, the movement and development of weather systems (*e.g.*, along their storm tracks across the oceans), and other processes (Wallace and Guzzler, 1981; Horel and Wallace, 1981; see Glantz *et al.*, 1991 for a review of the various mechanisms linking worldwide climate anomalies). The PNA and NAO patterns are now recognized as representing preferred structures of extratropical climate variations that are readily triggered by internal atmospheric mechanisms and also by surface boundary variations, especially from ocean sea surface temperatures (Hoskins and Karoly, 1981; Horel and Wallace, 1981; Simmons *et al.*, 1983).

As indicated in Chapter 2, these and other teleconnection patterns can be readily identified in the monthly and seasonal averages of atmospheric circulation anomalies in the free atmosphere using reanalysis data. Reanalysis data has also been instrumental in understanding the causes of teleconnection patterns and their North American surface climate impact (Feldstein 2000, 2002; Thompson and Wallace, 1998, 2000a,b). The ability to assess the relationships between teleconnections and their surface impacts provides an important foundation for attribution—North American climate variations are often due to particular atmospheric circulation patterns that connect climate anomalies over distant regions of the globe. Such a connection is illustrated schematically in Figure 3.1.

2) Attribution to surface boundary forcing
In some situations, teleconnections, including those described above, are a forced response to anomalous conditions at the Earth's surface. Under such circumstances, attribution statements that go beyond the statement of how recurrent features of the atmospheric circulation affect North American surface climate are feasible, and provide an explanation of the cause for the circulation itself. For instance, the atmospheric response to tropical Pacific sea surface temperature anomalies takes the form of a PNA-like pattern having significant impacts on North American climate, especially in the winter and spring seasons. However, other surface forcings, such as those related to

sea ice and soil moisture conditions, can also cause appreciable climate anomalies, although their influence is more local and does not usually involve teleconnections.

Bjerknes (1966, 1969) demonstrated that a surface pressure see-saw between the western and eastern tropical Pacific (now known as the Southern Oscillation) was linked with the occurrence of equatorial Pacific sea surface temperature (SST) anomalies, referred to as El Niño. This so-called El Niño-Southern Oscillation (ENSO) phenomenon was discovered to be an important source for year-to-year North American climate variation, with recent examples being the strong El Niño events of 1982 to 1983 and 1997 to 1998, whose major meteorological consequences over North America included flooding and storm damage over a wide portion of the western and southern United States and unusually warm winter temperatures over the northern United States (Rasmusson and Wallace, 1983). The cold phase of the cycle, referred to as La Niña, also has major impacts on North America, in particular, an enhanced drought risk across the southern and western United States (Ropelewski and Halpert, 1986; Cole *et al.*, 2002).

The impacts of ENSO on North American climate have been extensively documented using both historical data and sensitivity experiments in which the SST conditions associated with ENSO are specified in atmospheric climate models (see review by Trenberth *et al.*, 1998). Figure 3.2 illustrates the observed wintertime tropospheric circulation pattern during El Niño events of the last half century based on reanalysis data, and the associated North American surface signatures in temperature and precipitation. Reanalysis data is accurate enough to distinguish between the characteristic circulation pattern of the PNA (Figure 2.8) and that induced by ENSO—the latter having more widespread high pressure over Canada. Surface temperature features consist more of a north-south juxtaposition of warm-cold over North America during ENSO, as compared to the west-east structure associated with the PNA. The capacity to observe such distinctions is important when determining attribution because particular climate signatures indicate different possible causes.

North American climate variations are often due to particular atmospheric circulation patterns that connect climate anomalies over distant regions of the globe.

ENSO Impact

Temperature

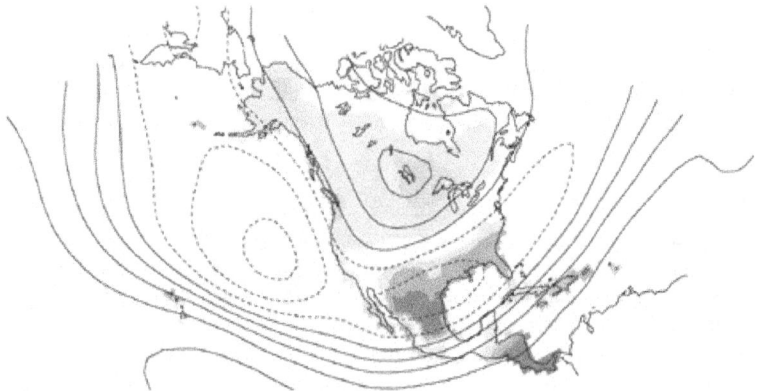

El Niño is a known
internal variation of
the coupled ocean-
atmosphere.

Precipitation

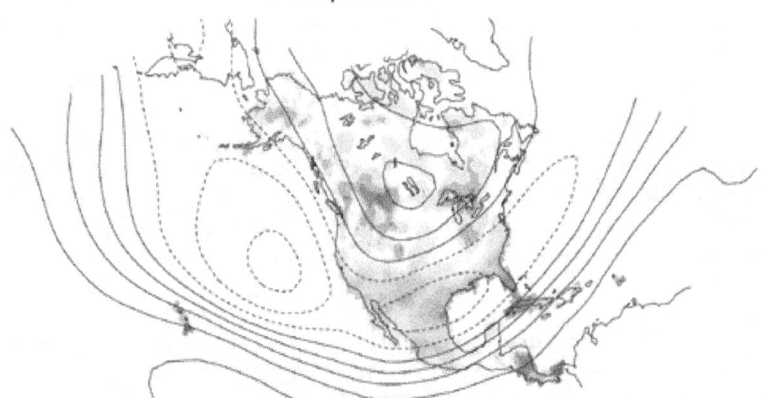

Figure 3.2 The correlation between a sea surface temperature index of El Niño-Southern Oscillation (ENSO) and 500 millibar (mb) pressure height field (contours). The shading indicates the correlations between ENSO index and the surface temperature (top panel) and the precipitation (bottom panel). The 500mb height is from the NCEP/NCAR RI reanalysis. The surface temperature and precipitation are from independent observational datasets. The correlations are based on seasonal mean winter (December-January-February) data for the period 1951 to 2006. The contours with negative correlation are dashed.

The use of climate models subjected to specified SSTs has been essential for determining the role of oceans in climate, and such tools are now extensively used in seasonal climate forecast practices. The atmospheric models are often subjected to realistic globally complete, monthly evolving SSTs (so-called AMIP experiments [Gates, 1992]) or to regionally confined idealized SST anomalies in order to explore specific cause-effect relations. These same models have also been used to assess the role of sea ice and soil moisture conditions on climate.

The process of forcing a climate model is discussed further in Box 3.2.

3) Attribution to external forcing
Explaining the origins for the surface boundary conditions themselves is another stage in attribution. El Niño, for example, is a known internal variation of the coupled ocean-atmosphere. On the other hand, a warming trend of ocean SST, as seen in recent decades over the tropical warm pool of the Indian Ocean/West Pacific, is recognized to result in part from changes in greenhouse gas forcing (Santer *et*

BOX 3.2: Forcing a Climate Model

The term "forcing" as used in Chapter 3 refers to a process for subjecting a climate model to a specified influence, often with the intention to probe cause-effect relationships. The imposed conditions could be "fixed" in time, such as a might be used to represent a sudden emission of aerosols by volcanic activity. It may be "time evolving" such as by specifying the history of sea surface temperature variations in an atmospheric model. The purpose of forcing a model is to study the Earth system response, and the degrees of freedom sensitivity of that response to both the model and the forcing employed. The schematic of the climate system helps to better understand the forcings used in various models of Chapter 3.

For atmospheric model simulations used in this SAP, the forcing consists of specified monthly evolving global sea surface temperatures during 1951 to 2006. By so restricting the lower boundary condition of the simulations, the response of unconstrained features of the climate system can be probed. In this SAP, the atmosphere and land surface are free to respond. Included in the former are the atmospheric hydrologic cycle involving clouds, precipitation, water vapor, temperature, and free atmospheric circulation. Included in the latter is soil moisture and snow cover, and changes in these can further feedback upon the atmosphere. Sea ice has been specified to climatological conditions in the simulations of this report, as has the chemical composition of the atmosphere including greenhouse gases, aerosols, and solar output.

For coupled ocean-atmosphere model simulations used in this SAP, the forcing consists of specified variations in atmospheric chemical composition (e.g,, carbon dioxide, methane, nitrous oxide), solar radiation, volcanic and anthropogenic aerosols. These are estimated from observations during 1951 to 2000, and then based upon a emissions scenario for 2001 to 2006. The atmosphere, land surface, ocean, and sea ice are free to respond to these specified conditions. The atmospheric response to those external forcings could result from the altered radiative forcing directly, though interactions and feedbacks involving the responses of the lower boundary conditions (e.g., oceans and cryosphere) are often of leading importance. For instance, much of the high-latitude amplification of surface air temperature warming due to greenhouse gas emissions is believed to result from such sea ice and snow cover feedback processes. Neither the coupled ocean-atmospheric models nor the atmospheric models used in this SAP include changes in land surface, vegetation, or ecosystems. Nor does the oceanic response in the coupled models include changes in biogeochemistry.

Multiple realizations of the climate models subjected to the same forcings are required in order to effectively separate the climate model's response from low-frequency climate variability. Ensemble methods are therefore used in Chapter 3. In the case of the atmospheric models, 33 total simulations (derived from two different models) forced as discussed above are studied. In the case of the coupled ocean-atmosphere models, 41 total simulations (derived from 19 different models) forced as discussed above are studied

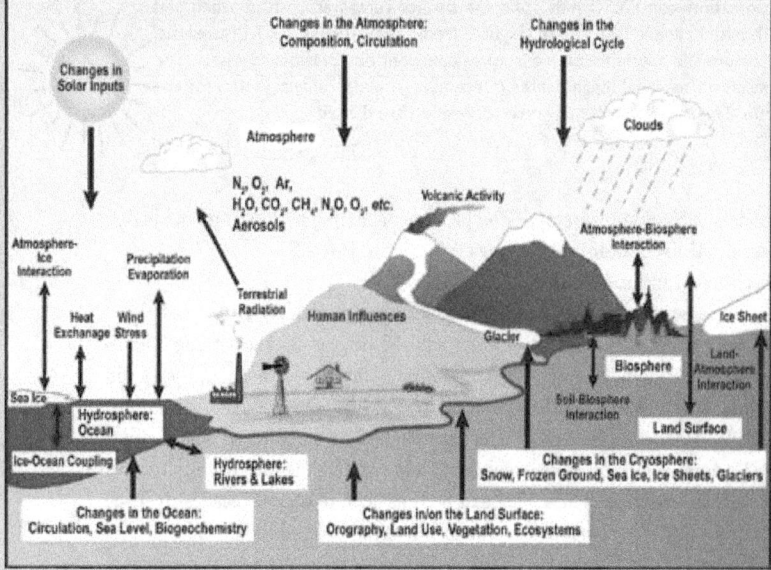

Figure Box 3.2 Schematic view of the components of the climate system, their processes and interactions (From IPCC, 2007a).

al., 2006; Knutson *et al.*, 2006). Figure 3.1 highlights the differences in how SSTs vary over the eastern *versus* western tropical Pacific as a consequence of different processes occurring in those regions. Thus, the remote effects of recent sea surface warming of the tropical ocean's warmest waters (the so-called warm pool) on North American climate might be judged to be of external origins to the ocean-atmosphere system, tied in part to changes in the atmosphere's chemical composition.

The third link in the attribution chain involves attribution of observed climate conditions to external forcing. The external forcing could be natural, for instance originating from volcanic aerosol effects or solar fluctuations, or the external forcing could be anthropogenic. As discussed extensively in the IPCC Reports, the attribution of climate conditions to external radiative forcing (greenhouse gases, solar, and volcanic forcing) can be done directly by specifying the natural and anthropogenic forcings within coupled ocean-atmosphere-land models. An indirect approach can also be used to attribute climate conditions to external forcing, for instance, probing the response of an atmospheric model to SST conditions believed to have been externally forced (Hoerling *et al.*, 2004). However, if an indirect approach is used, it can only be *qualitatively* determined that external forcing contributed to the event—an accurate *quantification* of the magnitude of the impact by external forcing can only be determined in a direct approach.

The tool used for attribution of external forcing, either to test the signal (see Section 3.1.2.2) due to anthropogenic greenhouse gas and atmospheric aerosol changes or land use changes, or natural external forcing due to volcanic and solar forcing, involves coupled ocean-atmosphere-land models forced by observed external forcing variations. As illustrated in Figure 3.1, this methodology has been widely used in the IPCC Reports to date. Several studies have used reanalysis data to first detect change in atmospheric circulation, and then test with models whether such change resulted from human influences. (Chapter 2 also discusses the use of reanalysis data in establishing the suitability of climate models used for attribution.) For instance, a trend in wintertime sea level

pressure has been observed and confirmed in reanalysis data that resembles the positive polarity of the NAO (high surface pressure over the midlatitude North Atlantic and low pressure over the Arctic), and greenhouse gas and sulfate aerosol changes due to human activities have been implicated as a contributing factor (Gillett *et al.*, 2003; Figure 3.7). Reanalysis data have been used to detect an increase in the height of the tropopause—a boundary separating the troposphere and stratosphere—and modeling results have established anthropogenic changes in stratospheric ozone and greenhouse gases as the primary cause (Santer *et al.*, 2003).

3.1.2.2 FINGERPRINTING

Many studies use climate models to predict the expected pattern of response to a forcing, referred to as "fingerprints" in the classic climate change literature, or more generally referred to as the "signal" (Mitchell *et al.*, 2001; IDAG, 2005; Hegerl *et al.*, 2007). The space and time scales used to analyze climate conditions are typically chosen so as to focus on the space and time scale of the signal itself, filtering out structure that is believed to be unrelated to forcing. For example, changes in greenhouse gas forcing are expected to cause a large-scale (global) pattern of warming that evolves slowly over time, and thus scientists often smooth data to remove small-scale variations in both time and space. On the other hand, it is expected that ENSO-related SST forcing yields a regionally focused pattern over the Pacific North American sector, having several centers of opposite signed anomalies (*i.e.*, warming or cooling), and therefore averaging over a large region such as this is inappropriate. To ensure that a strong signal has been derived from climate models, individual realizations of an ensemble, in which each member has been identically forced, are averaged. Ensemble methods are thus essential in separating the model's forced signal from its internal variability so as to minimize the mix of signal and noise, which results from unforced climatic fluctuations.

The consistency between an observed climate condition and the estimated response to a hypothesized key forcing is determined by (1) estimating the amplitude of the expected fingerprint empirically from observations; (2) assessing whether this estimate is statistically

Many studies use climate models to predict the expected pattern of response to a forcing, referred to as "fingerprints" in the classic climate change literature, or more generally referred to as the "signal".

consistent with the expected amplitude derived from forced model experiments; and then (3) inquiring whether the fingerprint related to the key forcing is distinguishable from that due to other forcings. The capability to do this also depends on the amplitude of the expected fingerprint relative to the noise.

In order to separate contributions by different forcings and to investigate if other combinations of forcing can also explain an observed event, the simultaneous effect of multiple forcings are also examined, typically using a statistical multiple regression analysis of observations onto several fingerprints representing climate responses to each forcing that, ideally, are clearly distinct from one another (Hasselmann, 1979; 1997; Allen and Tett, 1999; IDAG, 2005; Hegerl *et al.*, 2007). Examples include the known unique sign and global patterns of temperature response to increased sulfate aerosols (cooling of the troposphere, warming of the stratosphere) *versus* increased carbon dioxide (warming of the troposphere but cooling of the stratosphere). Another example is the known different spatial patterns of atmospheric circulation response over the North American region to SST forcing from the Indian Ocean compared to the tropical eastern Pacific Ocean (Simmons *et al.*, 1983; Barsugli and Sardeshmukh, 2002). If the climate responses to these key forcings can be distinguished, and if rescaled combinations of the responses to other forcings fail to explain the observed change, then the evidence for a causal connection is substantially increased. Thus, the attribution of recent large-scale warming to greenhouse gas forcing becomes more reliable if influences of other natural external forcings, such as solar variability, are explicitly accounted for in the analysis.

The confidence in attribution will thus be subject to the uncertainty in the fingerprints both estimated empirically from observations and numerically from forced model simulations. The effects of forcing uncertainties, which can be considerable for some forcing variables such as solar irradiance and aerosols, also remain difficult to evaluate despite recent advances in research.

Satellite and *in situ* observations during the reanalysis period yield reliable estimates of SST conditions over the world oceans, thus increasing the reliability of attribution based on SST forced atmospheric models. Estimates of other land surface conditions, including soil moisture and snow cover, are less reliable. Attribution results based on several models or several forcing observation histories also provide information on the effects of model and forcing uncertainty. Likewise, empirical estimates of fingerprints derived from various observational datasets provide information of uncertainty.

Finally, attribution requires knowledge of the internal climate variability on the time scales considered—the noise within the system against which the signal is to be detected and explained. The residual (remaining) variability in instrumental observations of the Earth system after the estimated effects of external forcing (*e.g.*, greenhouse gases and aerosols) have been removed is sometimes used to estimate internal variability of the coupled system. However, these observational estimates are uncertain because the instrumental records are too short to give a well-constrained estimate of internal variability, and because of uncertainties in the forcings and the corresponding estimates of responses. Thus, internal climate variability is usually estimated from long control simulations from climate models. Subsequently, an assessment is usually made of the consistency between the residual variability referred to above and the model-based estimates of internal variability; and analyses that yield implausibly large residuals are not considered credible. Confidence is further increased by comparisons between variability in observations and climate model data, by the ability of models to simulate modes of climate variability, and by comparisons between proxy reconstructions and climate simulations of the last millennium.

Sections 3.2, 3.3, 3.4, and 3.5 summarize current understanding on the causes of detected changes in North American climate. These sections will illustrate uses of reanalysis data in combination with surface temperature and precipitation measurements to examine the nature of North American climate variations, and compare with forced model experiments that test attributable causes. In addition, these sections also assess the current understanding of causes for other variations of significance

The attribution of recent large-scale warming to greenhouse gas forcing becomes more reliable if influences of other natural external forcings, such as solar variability, are explicitly accounted for in the analysis.

in North America's recent climate history, focusing especially on major North American droughts. In the mid-1930s, Congress requested that the Weather Bureau explain the causes for the 1930s Dust Bowl drought, with a key concern being to understand whether this event was more likely a multi-year occurrence or an indication of longer-term change. Similar to 70 years earlier, fundamental challenges in attribution science today are to distinguish quasi-cyclical variations from long-term trends, and natural from anthropogenic origins.

3.2 PRESENT UNDERSTANDING OF NORTH AMERICAN ANNUAL TEMPERATURE AND PRECIPITATION CLIMATE TRENDS FROM 1951 TO 2006

3.2.1 Summary of IPCC Fourth Assessment Report

Among the major findings of the IPCC Fourth Assessment Report (IPCC, 2007b) is that "it is *likely* that there has been significant anthropogenic warming over the past 50 years averaged over each continent except Antarctica". This conclusion was based on recent fingerprint-based studies on the attribution of annual surface temperature involving space-time patterns of temperature variations and trends. Model studies using only natural external forcings were unable to explain the warming over North America in recent decades, and only experiments including the effects of anthropogenic forcings reproduced the recent upward trend. The IPCC Report also stated that, for precipitation, there was low confidence in detecting and attributing a change, especially at the regional level.

This assessment focuses in greater detail on North American temperature and precipitation variability during the period 1951 to 2006.

The origins for the North American fluctuations are assessed by examining the impacts on North America from time-evolving sea surface conditions (including ENSO and decadal ocean variations), in addition to time evolving anthropogenic effects. The use of reanalysis data to aid in the attribution of surface climate conditions is illustrated.

3.2.2 North American Annual Mean Temperature

3.2.2.1 DESCRIPTION OF THE OBSERVED VARIABILITY

Seven of the warmest ten years since 1951 have occurred in the last decade (1997 to 2006). The manner in which North American annual temperatures have risen since 1951, however, has been neither smooth nor consistent, being characterized by occasional peaks and valleys (Figure 3.3, top). The coldest year since 1951 occurred in 1972, and below average annual temperatures occurred as recently as 1996. Explanations for such substantial variability are no less important than explanations for the warming trend.

Virtually all of the warming averaged over North America since 1951 has occurred after 1970. It is noteworthy that North American temperatures cooled during the period 1951 through the early 1970s. In the 1970s, the general public and policy makers were interested in finding the reason for this cooling, with concerns about food production and societal disruptions. They turned to the meteorological community for expert assessment. Unfortunately, climate science was in its early stages and attribution was considerably more art than science. The essential tools for performing rigorous attribution such as global climate models were not yet available, nor was much known then about the range of historical climate variations such as those that have been subsequently revealed by paleoclimate studies. A consistent climate analysis of the historical instrumental record that included descriptions of the free atmosphere was also unavailable.

Barring an explanation of the cause for the cooling, and with no comprehensive climate models available, some scientists responded to the public inquiries on what would happen by merely extrapolating recent trends, thereby portraying an increased risk for a cooling world (Kukla and Mathews, 1972). Others suggested in the mid-1970s that we might be at the brink of a pronounced global warming, arguing that internal variations of the climate were then masking an anthropogenic signal (Broecker, 1975). The 1975 National Academy of Sciences report (NRC, 1975) on understanding climate change emphasized the fragmentary state of knowledge

Seven of the warmest ten years since 1951 have occurred in the last decade (1997 to 2006). Virtually all of the warming averaged over North America since 1951 has occurred after 1970.

BOX 3.3: Choosing the Assessment Period

The authors of this Product were asked to examine the strengths and limitations of current reanalysis products, and to assess capabilities for attributing the causes for climate variations and trends during the reanalysis period. The scope of this assessment is thus bounded by the reanalysis record (1948 to present). An important further consideration is the availability of sufficient, quality controlled surface observations to define key climate variations accurately. For precipitation, a high quality global gridded analysis is available beginning in 1951, thereby focusing the attribution to the period from 1951 to 2006.

It is reasonable to ask whether such a 56-year assessment period adequately samples the principal features of climate variability. Does it, for example, capture the major climate events, such as droughts, that may be of particular concern to decision makers? Is it a sufficiently long period to permit the distinction between fluctuations in climate conditions that are transient, or cyclical, from trends that are related to a changing climate? How well do scientists understand the climate conditions prior to 1951, and what insight does analysis of those conditions provide toward explaining post-1950 conditions? These are all important questions to bear in mind when reading this Product, especially if one wishes to generalize conclusions about the nature of and causes for climate conditions during 1951 to 2006 to earlier or future periods.

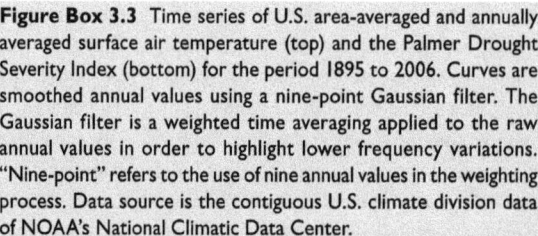

As a case in point, the U.S. surface temperature record since 1895 is remarkable for its multi-decadal fluctuations (top panel). A simple linear trend fails to describe all features of U.S. climate variations, and furthermore, a trend analysis for any subset of this 112-year period may be problematic since it may capture merely a segment of a transient oscillation. The 1930s was a particularly warm period, one only recently eclipsed. The United States has undergone two major swings between cold epochs (beginning in

Figure Box 3.3 Time series of U.S. area-averaged and annually averaged surface air temperature (top) and the Palmer Drought Severity Index (bottom) for the period 1895 to 2006. Curves are smoothed annual values using a nine-point Gaussian filter. The Gaussian filter is a weighted time averaging applied to the raw annual values in order to highlight lower frequency variations. "Nine-point" refers to the use of nine annual values in the weighting process. Data source is the contiguous U.S. climate division data of NOAA's National Climatic Data Center.

the 1890s and 1960s) and warm epochs (1930s and 2000s). It is reasonable to wonder whether the current warmth will also revert to colder conditions in coming decades akin to events following the 1930s peak, and attribution science is therefore important for determining whether the same factors are responsible for both warmings or not. Some studies reveal that the earlier warming may have resulted from a combination of anthropogenic forcing and an unusually large natural multi-decadal fluctuation of climate (Delworth and Knutson, 2000). Other work indicates a contribution to the early twentieth century warming by natural forcing of climate, such as changes in solar radiation or volcanic activity (e.g., Hegerl et al., 2006). The 1930s warming was part of a warming focused mainly in the northern high latitudes, a pattern reminiscent of an increase in poleward ocean heat transport (Rind and Chandler, 1991), which can itself be looked upon as due to "natural variability". In contrast, the recent warming is part of a global increase in temperatures, and the IPCC Fourth Assessment Report, Chapter 9 states that it is likely that a significant part of warming over the past 50 years over North America may be human related (IPCC, 2007a), thus contrasting causes of the warming that occurred in this period from that in 1930s. The physical processes related to this recent warming are further examined in this Chapter.

The year 1934 continues to stand out as one the warmest years in the United States' 112-year record, while averaged over the entire globe, 1934 is considerably cooler than the recent decade. The warmth of the United States in the 1930s coincided with the Dust Bowl (lower panel), and drought conditions likely played a major role in increasing land surface temperatures. Prior studies suggest that the low precipitation during the Dust Bowl was related in part to sea surface temperature conditions over the tropical oceans (Schubert et al., 2004; Seager et al., 2005). Current understanding of severe U.S. droughts that have occurred during the reanalysis period as described in this Chapter builds upon such studies of the Dust Bowl.

of the mechanisms causing climate variations and change, and posed the question of whether scientists would be able to recognize the first phases of a truly significant climate change when it does occur (NRC, 1975). Perhaps the single most important attribution challenge today regarding the time series of Figure 3.3 is whether the reversal of the cooling trend after 1975 represents such a change, and one for which a causal explanation can be offered.

There is very high confidence in the detection that the observed temperature trend reversed after the early 1970s. The shaded area in Figure 3.3 (top right panel) illustrates the spread among four different analyses of surface measurements (see Table 3.2 for descriptions of these data), and the analysis uncertainty as revealed by their range is small compared to the amplitude of the trend and principal variations. Also shown is the surface temperature time series derived from the reanalysis. Despite the fact that the assimilating model used in producing the NCEP/NCAR reanalysis does not incorporate observations of surface temperature (Kalnay *et al.*, 1996), the agreement with the *in situ* observations is strong. This indicates that the surface temperature averaged over the large domain of North America is constrained by and is consistent with climate conditions in the free atmosphere. Both for the emergent warming trend in the 1970s, and for the variations about it, this excellent agreement among time series based on different observational datasets and the reanalysis increases confidence that they are not artifacts of analysis procedure.

The total 1951 to 2006 change in observed North American annual surface temperatures is +0.90°C ± 0.1°C (about +1.6°F ± 0.2°F), with the uncertainty estimated from the range between trends derived from four different observational analyses. Has a *significant* North American warming been detected? Answers to this question require knowledge of the plausible range

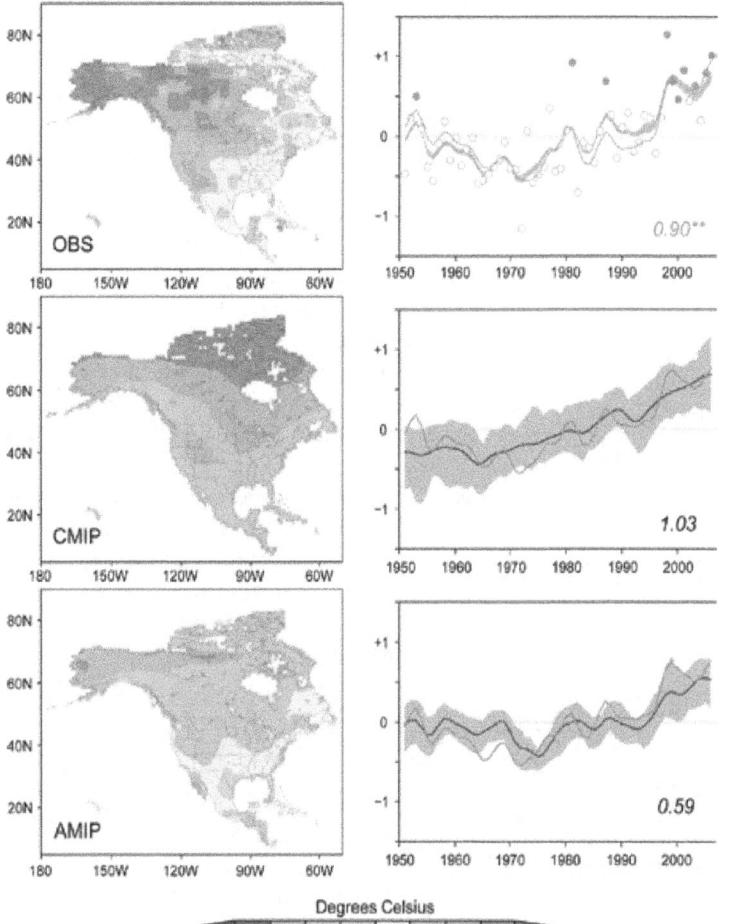

North America Annual Temperature: 1951-2006

in 56-year trends that can occur naturally in the absence of any time varying anthropogenic forcing. The length of the observational record does not permit such an assessment, but an analysis of such variations in coupled model simulations that exclude variations in anthropogenic forcing provides an indirect estimate. To estimate the confidence that a change in

Figure 3.3 The 1951 to 2006 trend in annually averaged North American surface temperature from observations (top), CMIP simulations (middle), AMIP simulations (bottom). Maps (left side) show the linear trend in annual temperatures for 1951 to 2006 (units, °C change over 56 years). Time series (right side) show the annual values from 1951 to 2006 of surface temperatures averaged over the whole of North America. Curves are smoothed annual values using a five-point Gaussian filter, based on the average of four gridded surface observational analyses, and the ensemble mean of climate simulations. The Gaussian filter is a weighted time-averaging applied to the raw annual values in order to highlight lower frequency variations. "Five-point" refers to the use of five annual values in the weighting process. Unsmoothed annual observed temperatures are shown by red circles, with filled circles denoting the ten warmest years since 1951. Plotted values are the total 56-year change (°C), with the double asterisks denoting very high confidence that an observed change was detected. For observations, the gray band denotes the range among four surface temperature analyses. The blue curve is the NCEP/NCAR reanalysis surface temperature time series. For simulations, the gray band contains the 5 to 95 percent occurrence of individual model simulations.

Numerous detection and attribution studies have shown that the observed warming of North American surface temperature since 1950 cannot be explained by natural climate variations. alone and is consistent with the response to anthropogenic climate forcing, particularly increases in atmospheric greenhouse gases

North American temperatures has been detected, a non-parametric test, which makes no assumptions about the statistical distribution of the data, has been applied that estimates the range of 56-year trends attributable to natural variability alone (see Appendix B for methodological details). A diagnosis of 56-year trends from the suite of "naturally forced" Coupled Model Intercomparison Project (CMIP) runs is performed, from which a sample of 76 such trends were generated for annual North American average surface temperatures. Of these 76 "trends estimates" consistent with natural variability, no single estimate was found to generate a 56-year trend as large as observed.

It is thus *very likely* that a change in North American annual mean surface temperature has been detected. That assessment takes into account the realization that the climate models have biases that can affect statistics of their simulated internal climate variability.

3.2.2.2 ATTRIBUTION OF THE OBSERVED VARIATIONS

3.2.2.2.1 External Forcing

The IPCC Fourth Assessment Report provided strong attribution evidence for a significant anthropogenic greenhouse gas forced warming of North American surface temperatures (IPCC, 2007a). Figure 3.4 is drawn from that Report, and compares continental-averaged surface temperature changes observed with those simulated using the CMIP coupled models having both natural and anthropogenic forcing. It is clear that only experiments using observed time varying anthropogenic forcing explain the warming in recent decades. Numerous detection and attribution studies, as reviewed by Hegerl *et al.* (2007), have shown that the observed warming of North American surface temperature since 1950 cannot be explained by natural climate variations alone and is consistent with the response to anthropogenic climate forcing, particularly increases in atmospheric greenhouse gases (Karoly *et al.*, 2003; Stott, 2003; Zwiers and Zhang, 2003; Knutson *et al.*, 2006; Zhang *et al.*, 2006). The suitability of these coupled climate models for attribution is indicated by the fact that they are able to simulate variability on time scales of decades and longer that is consistent with reanalysis data of the free atmosphere and surface observations over North America (Hegerl *et al.*, 2007).

A more detailed examination of the human influence on North America is provided in Figure 3.3 (middle) that shows the spatial map of the 1951 to 2006 model-simulated surface temperature trend, in addition to the trend over time. There are several key agreements between the CMIP simulations and observations that support the argument for an anthropogenic effect. First, both indicate that most of the warming has occurred in the past 30 years. The North American warming after 1970 is thus *likely* the result of the region's response to anthropogenic forcing. Second, the total 1951 to 2006 change in observed North American annual surface tem-

Figure 3.4 Temperature changes relative to the corresponding temperature average for 1901 to 1950 (°C) from decade to decade for the period 1906 to 2005 over the Earth's continents, as well as the entire globe, global land area, and the global ocean (lower graphs). The black line indicates observed temperature change and the colored bands show the combined range covered by 90 percent of recent model simulations. Red indicates simulations that include natural and human factors, while blue indicates simulations that include only natural factors. Dashed black lines indicate decades and continental regions for which there are substantially fewer observations compared with other continents during that time. Detailed descriptions of this figure and the methodology used in its production are given in Hegerl *et al.* (2007).

peratures of +0.90°C (about +1.6°F) compares well to the simulated ensemble averaged warming of +1.03°C (almost +1.9°F). Whereas the observed 56-year trend was shown in Section 3.2.2.1 to be inconsistent with the population of trends drawn from a state of natural climate variability, the observed warming is found to be consistent with the population of trends drawn from a state that includes observed changes in the anthropogenic greenhouse gas forcing during 1951 to 2006.

Further, the observed low frequency variations of annual temperature fall within the 5 to 95 percent uncertainty range of the individual model simulations. All CMIP runs that include anthropogenic forcing produce a North American warming during 1951 to 2006. For some simulations, the trend is less than that observed and for some it is greater than that observed. This range results from two main factors. One is the uncertainty in anthropogenic signals; namely that the individual 19 models subjected to identical forcing exhibit somewhat different sensitivities. The other results from the internal variability of the models; namely that individual runs of the same model generate a

range of anomalies owing to natural coupled-ocean atmosphere fluctuations.

Each of the 41 anthropogenic forced simulations produces a 56-year North American warming (1951 to 2006) that accounts for more than half of the observed warming. Our assessment of the origin for the observed North American surface temperature trend is that more than half of the warming during 1951 to 2006 is *likely* the result of human influences. It is exceptionally *unlikely* that the observed warming has resulted from natural variability alone because there is a clear separation between the ensembles of climate model simulations that include only natural forcings and those that contain both anthropogenic and natural forcings (Hegerl *et al.*, 2007). These confidence statements reflect the uncertainty of the role played by model biases in their sensitivity to external forcing, and also the unknown impact of biases on the range of their unforced natural variability.

From Figure 3.3, it is evident that the yearly fluctuations in observed North American temperature are of greater amplitude than those occurring in the ensemble average of externally

> More than half of the warming over North America during 1951 to 2006 is *likely* the result of human influences. It is exceptionally *unlikely* that the observed warming has resulted from natural variability alone.

BOX 3.4: Use of Expert Assessment

The use of expert assessment is a necessary element in attribution as a means to treat the complexities that generate uncertainties. Expert assessment is used to define levels of confidence, and the terms used in this Product (see Preface) follow those of the IPCC Fourth Assessment Report. The attribution statements used in Chapter 3 of this SAP also employ probabilistic language (for example, "virtually certain") to indicate a likelihood of occurrence.

To appreciate the need for expert assessment, it is useful to highlight the sources of uncertainty that arise in seeking the cause for climate conditions. The scientific process of attribution involves various tools to probe cause-effect relationships such as historical observations, climate system models, and mechanistic theoretical models. Despite ongoing improvements in reanalysis and models, these and other tools have inherent biases rendering explanations of the cause for a climate condition uncertain. Uncertainty can arise in determining a forced signal (*i.e.*, fingerprint identification). For instance, the aerosol-induced climate signal involves direct radiative effects that require on accurate knowledge of the amount and distribution of aerosols in the atmosphere. These are not well observed quantities, leading to so-called "value uncertainties" (IPCC, 2007a) because the forcing itself is poorly known. The aerosol-induced signal also involves an indirect radiative forcing, the latter depending on cloud properties and water droplet distributions. These cloud radiative interactions are poorly represented in current generation climate models (Kiehl, 1999), contributing to so-called "structural uncertainties" (IPCC, 2007a). Even if the forcing is known precisely and the model includes the relevant processes and relationships, the induced signal may be difficult to distinguish from other patterns of climate variability thereby confounding the attribution.

The scientific peer-reviewed literature provides a valuable guide to the author team of Chapter 3 for determining attribution confidence. In addition, new analyses in this Product are also examined in order to provide additional information. These employ methods and techniques that have been extensively tested and used in the scientific literature. In most cases, new analyses involve observational data and model simulations that have merely updated to include recent years through 2006.

The oceans play a major role in climate, not only for determining its average conditions and seasonal cycle, but also for determining its anomalous conditions including interannual-to-decadal fluctuations.

forced runs. This is consistent with the fact that the observations blend the effects of internal and external influences while the model estimates only the time-evolving impact of external forcings. Nonetheless, several of these observed fluctuations align well with those in the CMIP data. In particular, the model warming trend is at times punctuated by short periods of cooling, and these episodes coincide with major tropical volcanic eruptions (*e.g.*, Aguang in 1963; Mt. Pinatubo in 1991). These natural externally forced cooling episodes correspond well with periods of observed cooling, as will be discussed further in Section 3.4.

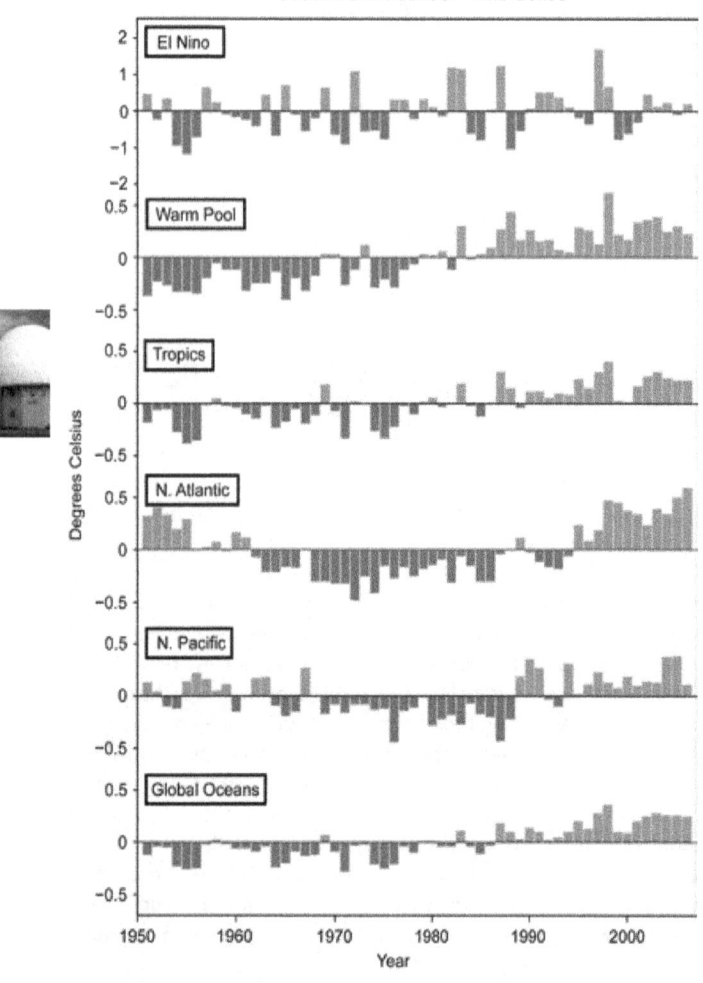

Figure 3.5 Observed annual mean sea surface temperature (SST) time series for 1951 to 2006. The oceanic regions used to compute the indices are 5°N to 5°S, 90°W to 150°W for El Niño, 10°S to 10°N, 60°E to 150°E for the warm pool, 30°S to 30°N for the tropics, 30°N to 60°N for the North Atlantic, 30°N to 60°N for the North Pacific, and 40°S to 60°N for the global oceans. The dataset used is the HadiSST monthly gridded fields, and anomalies are calculated relative to a 1951 to 2006 climatological reference.

3.2.2.2.2 Sea Surface Temperature Forcing
The oceans play a major role in climate, not only for determining its average conditions and seasonal cycle, but also for determining its anomalous conditions including interannual-to-decadal fluctuations. Section 3.1 discussed modes of anomalous sea surface temperature (SST) variations that impact North America, in particular those associated with ENSO. Figure 3.5 illustrates the variations in time of SSTs over the global oceans and over various ocean basins during 1951 to 2006. Three characteristic features of the observed SST fluctuations are noteworthy. First, SSTs in the eastern tropical Pacific (top panel) vary strongly from year to year, as warm events alternate with cold events, which is indicative of the ENSO cycle. Second, extratropical North Pacific and North Atlantic SSTs have strong year-to-year persistence, with decadal periods of cold conditions followed by decadal periods of warm conditions. Third, the warm pool of the tropical Indian Ocean/West Pacific, the tropically averaged SSTs, and globally averaged SSTs are dominated by a warming trend. In many ways, these resemble the North American surface temperature changes over time, including a fairly rapid emergence of warmth after the 1970s.

A common tool for determining the SST effects on climate is the use of atmospheric general circulation models (AGCM) forced with the specified time evolution of the observed SSTs, in addition to empirical methodologies (see Figure 3.2 for the El Niño impact inferred from reanalysis data, and Box 3.1 for an assessment of model simulated ENSO teleconnections). Such numerical modeling approaches are generally referred to as AMIP simulations (Gates, 1992), and that term is adapted in this Product to refer to model runs spanning the period 1951 to 2006.

Much of the known effect of SSTs has focused on the boreal winter season, a time when ENSO and its impacts on North America are at their peak. However, the influence of SSTs on annual average variability over North America is not yet documented in the peer-reviewed literature. Therefore, an expert assessment is presented in this Section based on the analysis of two AGCMs (Table 3.1). It is important to note that the AMIP simulations used in this analysis do not include the observed evolution of external forcings (*e.g.*, solar, volcanic aerosols, or anthropogenic greenhouse gases). The specified SSTs may, however, reflect the footprints of such external influences (see Section 3.4 and Figure 3.18 for a discussion of the same SST time series constructed from the CMIP simulations).

North American annual temperature trends and their evolution over time are well replicated in the AMIP simulations (Figure 3.3, bottom). There are several key agreements between the AMIP simulations and observations that support the argument for an SST effect. First, most of the AMIP simulated warming occurs after 1970, in agreement with observations. The time evolution of simulated annual North American surface temperature fluctuations is very realistic, with a temporal correlation of 0.79 between the raw unsmoothed observed data and simulated annual values. While slightly greater than the observed correlation with CMIP of 0.68, much of the positive year-over-year correlation is due to the warming trend. Second, the AMIP pattern correlation of 0.87 with the observed trend map highlights the remarkable spatial agreement and exceeds the 0.79 spatial correlation for the CMIP simulated trend. Several other notable features of the AMIP simulations include the greater warming over western North America and slight cooling over eastern and southern regions of the United States. The total 1951 to 2006 change in observed North American annual surface temperatures of +0.90°C (about 1.6°F) compares well to the AMIP simulated warming of +0.59°C (almost 1.1°F).

A strong agreement exists between the AMIP and CMIP simulated North American surface temperature trend patterns and their time evolutions during 1951 to 2006. This comparison of the CMIP and AMIP simulations indicates that a substantial fraction of the area-averaged anthropogenic warming over North America has *likely* occurred as a consequence of sea surface temperature forcing. However, the physical processes by which the oceans have led to North American warming is not currently known.

An important attribution challenge is determining which aspects of regional SST variability during 1951 to 2006 have been important in contributing to the signals shown in Figure 3.3. Idealized studies linking regional SST anomalies to atmospheric variability have been conducted (Hoerling *et al.*, 2001; Robertson *et al.*, 2003; Barsugli *et al.*, 2002; Kushnir *et al.*, 2002); however, a comprehensive suite of model simulations to address variability in North American surface temperatures during 1951 to 2006 has not yet been undertaken. Whereas the North American sensitivity to SST forcing from the ENSO region is well understood, the effect of the progressive tropical-wide SST warming, a condition that has been the major driver of globally averaged SST behavior during the last half century, is less well known (Figure 3.5). A further question is the effect that recent decadal warming of the North Pacific and North Atlantic Oceans have had on North American climate, either in explaining the spatial variations in North American temperature trends or as a factor in the accelerated pace of North American warming since 1970. Although the desired simulation suite have yet to be conducted, some attribution evidence for regional SST effects can be learned empirically from the reanalysis data itself, which are capable of describing changes in tropospheric circulation patterns, elements of which are known to have regional SST sources. This will be the subject of further discussion in Section 3.3, where observed changes in PNA and NAO circulation patterns since 1950 are described and their role in North American climate trends is assessed.

3.2.2.2.3 Analysis of Annual Average Rainfall Variability Over North America

North American precipitation exhibits considerably greater variability in both space and time compared with temperature. The annual cycle of precipitation varies greatly across the continent, with maximum winter amounts along western North America, maximum summertime amounts over Mexico and Central America, and comparatively little seasonality over eastern North America. Therefore, it is not surprising that the 1951 to 2006 trends in annual precipitation are mainly regional in nature (Figure 3.6, top). Several of these trends are discussed further in Section 3.3.

For area-averaged North America as a whole, there is no coherent trend in observed precipitation since 1951. The time series of annual values has varied within 10 percent of the 56-year

North American precipitation exhibits considerably greater variability in both space and time compared with temperature.

North American Annual Precipitation: 1951-2006

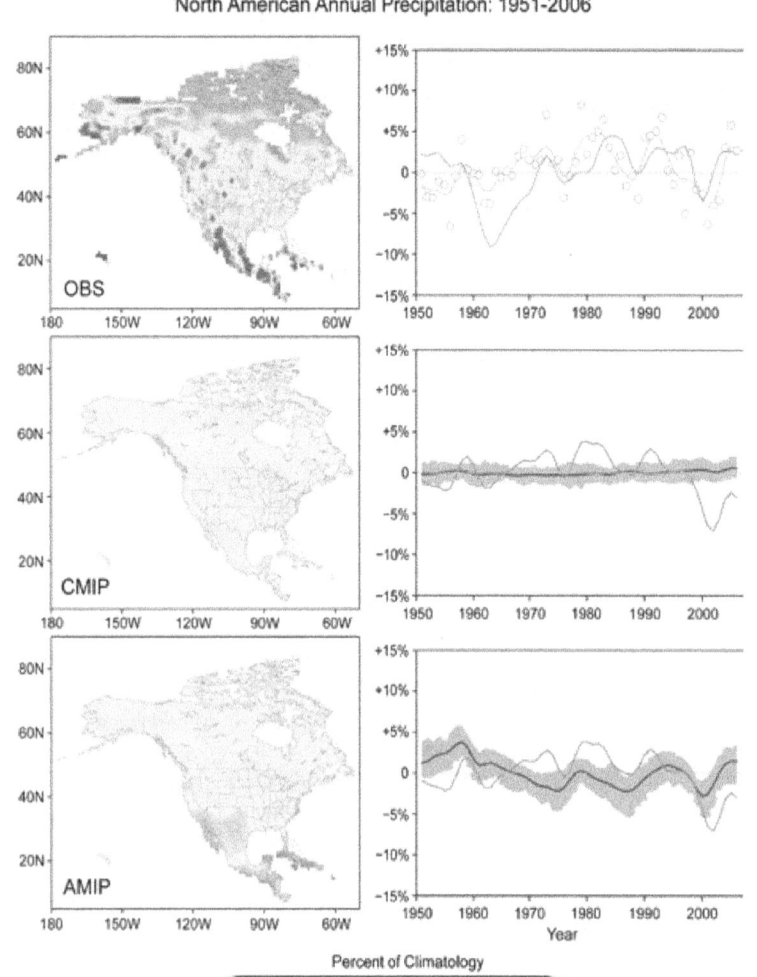

Percent of Climatology

-50   -25   -15   0   15   25   50

Figure 3.6 The 1951 to 2006 trend in annually averaged North American precipitation from observations (top), CMIP simulations (middle), AMIP simulations (bottom). Maps (left side) show the linear trend in annual precipitations for 1951 to 2006 (units, total 56-year change as percent of the climatological average). Time series (right side) show the annual values from 1951 to 2006 compared as a percentage of the 56-year climatological precipitation average. Curves are smoothed annual values using a five-point Gaussian filter, based on the Global Precipitation Climatology Center observational analysis, and the ensemble mean of climate simulations. The Gaussian filter is a weighted time averaging applied to the raw annual values in order to highlight lower frequency variations. "Five-point" refers to the use of five annual values in the weighting process. Unsmoothed annual observed precipitation is shown by red circles. The blue curve is the NCEP/NCAR reanalysis precipitation over time. For simulations, the gray band contains the 5 to 95 percent occurrence of individual model simulations.

between different types of anomalies when averaging across the continent, as is done in Figure 3.6. For instance, above average precipitation due to excess rain in one region can offset below average precipitation due to drought in another region.

Neither externally forced nor SST forced simulations show a significant change in North American-wide precipitation since 1951. In addition, the area-averaged annual fluctuations in the simulations are generally within a few percent of the 56-year climatological average (Figure 3.6, middle and bottom panels). The comparison of the observed and CMIP simulated North America precipitation indicates that the anthropogenic signal is small relative to the observed variability over years and decades. As a note of caution regarding the suitability of the CMIP models for precipitation, the time series of North American precipitation in the individual CMIP simulations show much weaker decadal variability than is observed. Note especially that the recent observed dry anomalies reside well outside the range of outcomes produced by all available CMIP runs, suggesting that the models may underestimate the observed variability, at least for North American annual and area averages.

A small number of detection and attribution studies of average precipitation over land have identified a signal due to volcanic aerosols in low frequency variations of precipitation (Gillett *et al.*, 2004; Lambert *et al.*, 2004). Climate models appear to underestimate both the variation of average precipitation over land compared to

climatological precipitation average, with the most notable feature being the cluster of dry years from the late 1990s to the early 2000s. However, even these annual variations for North American averaged precipitation as a whole are of uncertain physical significance because of the regional focus of precipitation fluctuations and the considerable cancellation

observations and the observed precipitation changes in response to volcanic eruptions (Gillett *et al.*, 2004; Lambert *et al.*, 2004). Zhang *et al.* (2007) examined the human influence on precipitation trends over land within latitudinal bands during 1950 to 1999, finding evidence for anthropogenic drying in the subtropics and increased precipitation over sub-polar latitudes, though observed and greenhouse gas forced simulations disagreed over much of North America.

The time series of North America precipitation from the AMIP simulations shows better agreement with the observations than the CMIP simulations, including marked negative anomalies (*e.g.*, droughts) over the last decade. This suggests that a part of the observed low frequency variations stems from observed variations of global SST. A connection between ENSO-related tropical SST anomalies and rainfall variability over North America has been well documented, particularly for the boreal winter, as mentioned earlier. In addition, the recent years of dryness are consistent with the multi-year occurrence of La Niña (Figure 3.5). The influence of tropical-wide SSTs and droughts in the midlatitudes and North America has also been documented in previous studies (Hoerling and Kumar, 2003; Schubert *et al.*, 2004; Lau *et al.*, 2006; Seager *et al.*, 2005; Herweijer *et al.*, 2006). Such causal links do provide an explanation for the success of AMIP integrations in simulating and explaining some aspects of the observed variability in North American area-averaged precipitation, although it is again important to recognize the limited value of such an area average for describing moisture related climate variations.

3.3 PRESENT UNDERSTANDING OF UNITED STATES SEASONAL AND REGIONAL DIFFERENCES IN TEMPERATURE AND PRECIPITATION TRENDS FROM 1951 TO 2006

3.3.1 Introduction

As noted in the recent IPCC Fourth Assessment Report, identification of human causes for variations or trends in temperature and precipitation at regional and seasonal scales is more difficult than for larger area and annual averages (IPCC, 2007a). The primary reason is that internal climate variability is greater at these scales—averaging over larger space-time scales reduces the magnitude of the internal climate variations (Hegerl *et al.*, 2007). Early idealized studies (Stott and Tett, 1998) indicated that the spatial variations of surface temperature changes due to changes in external forcing, such as green-

house gas related forcings, would be detectable only at scales of 5000 kilometers (about 3100 miles) or more. However, these signals will be more easily detectable as the magnitude of the expected forced response increases with time. The IPCC Fourth Assessment Report highlights the acceleration of the warming response in recent decades (IPCC, 2007a).

Consistent with increased external forcing in recent decades, several studies (Karoly and Wu, 2005; Knutson *et al.*, 2006; Wu and Karoly, 2007; Hoerling *et al.*, 2007) have shown that the warming trends over the second half of the twentieth century at many individual cells, which are 5° latitude by 5° longitude in area (about 556 by 417 kilometers or 345 by 259 miles), across the globe can now be detected in observations. Further, these are also consistent with the modeled response to anthropogenic climate forcing and cannot be explained by internal variability and response to natural external forcing alone. However, there are a number of regions that do not show significant warming, including the southeast United States, although modeling results have yet to consider a range of other possible forcing factors that may be more important at regional scales, including changes in carbonaceous aerosols (IPCC, 2007a) and changes in land use and land cover (Pielke *et al.*, 2002; McPherson, 2007).

What is the current capability to explain spatial variations and seasonal differences in North American climate trends over the past half-century? Can various differences in space and time be accounted for by the climate system's sensitivity to time evolving anthropogenic forcing? To what extent can the influences of natural processes be identified? Recent studies have linked some regional and seasonal variations in temperature and precipitation over the United States to variations in SST (*e.g.*, Livezey *et al.*, 1997; Kumar *et al.*, 2001; Hoerling and Kumar 2002; Schubert *et al.*, 2004; Seager *et al.*, 2005). These published results have either focused on annually averaged or winter-only conditions. This Product will assess both the winter and summer origins change over North America, the contiguous United States, and various sub-regions of the United States.

Recent studies have linked some regional and seasonal variations in temperature and precipitation over the United States to variations in sea surface temperature.

3.3.2 Temperature Trends

3.3.2.1 NORTH AMERICA

The observed annually averaged temperature trends over North America in Figure 3.3 show considerable variation in space, with the largest warming over northern and western North America and least warming over the southeastern United States. The ensemble-averaged model response to anthropogenic and natural forcing since 1951 (CMIP runs in Figure 3.3) shows a more uniform warming pattern, with larger values in higher latitudes and in the interior of the continent. While the spatial correlation of the CMIP simulations with observations for the 1951 to 2006 North American surface temperature trend is 0.79, that agreement is almost entirely due to the agreement in the area-averaged temperature trend. Upon removing the area-averaged warming, a process that highlights the spatial variations, the resulting pattern correlation between trends in CMIP and observations is only 0.13. Thus, the spatial variations in observed North American surface temperature change since 1951 are *unlikely* to be due to anthropogenic forcing alone.

An assessment of AMIP simulations indicates that key features of the spatial variations of annually averaged temperature trends are more consistent with a response to SST variations during 1951 to 2006. The ensemble-averaged model response to observed SST variations (CMIP runs in Figure 3.3) shows a spatial pattern of North American surface temperature trends that agrees well with the observed pattern, with a correlation of 0.87. Upon removing the area-averaged warming, the resulting correlation is still 0.57. This indicates that the spatial variation of the observed warming over North America is *likely* influenced by observed regional SST variations, which is consistent with the previously published results of Robinson *et al.* (2002) and Kunkel *et al.* (2006).

A diagnosis of observed trends in free atmospheric circulation, using the reanalysis data of 500 millibar (mb) pressure heights, provides a physical basis for the observed regionality in North American surface temperature trends. Figure 3.7 illustrates the 1951 to 2006 November to April surface temperature trends together with the superimposed 500 mb height trends. It is during the cold half of the year that many of the spatial features in the annual trend originate, a time during which teleconnection patterns are also best developed and exert their strongest impacts. The reanalysis data captures two prominent features of

> The spatial variations in observed North American surface temperature change since 1951 are *unlikely* to be due to anthropogenic forcing alone. The spatial variation of the observed warming over North America is *likely* influenced by observed regional sea surface temperature variations.

North American Winter
Circulation and Temperature Change

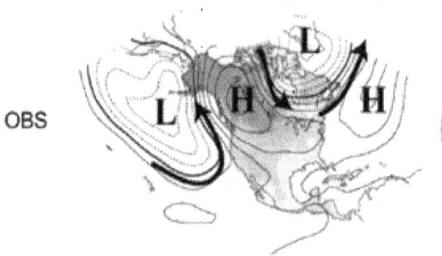

OBS

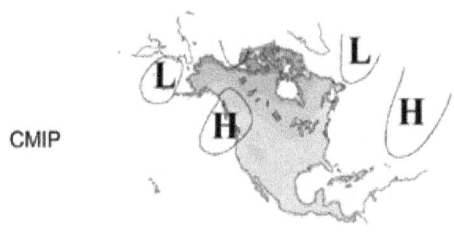

CMIP

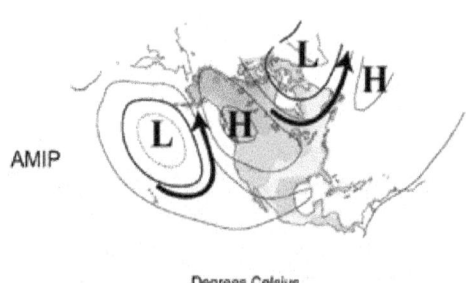

AMIP

Figure 3.7 The 1951 to 2006 November to April trend of 500 millibars heights (contours, units meters total change over 56 year period, contour interval 10 meters) and North American surface temperature (color shading, units °C total change over 56 year period) for observations (top), CMIP ensemble mean (middle), AMIP ensemble mean (bottom). Anomalous High and Low Pressure regions are highlighted. Arrows indicate the anomalous wind direction, which circulates around the High (Low) Pressure centers in a clockwise (counterclockwise) direction.

circulation change since 1951, one that projects upon the positive phase of the Pacific North American pattern and the other that projects upon the positive phase of the North Atlantic Oscillation pattern. Recalling from Chapter 2 the surface temperature fingerprints attributable to the PNA and NAO, the diagnosis in Figure 3.7 reveals that the pattern of observed surface temperature trend can be understood as a linear combination of two separate physical patterns, consistent with prior published results of Hurrell (1995) and Hurrell (1996).

The historical reanalysis data thus proves invaluable for providing a physically consistent description of the regional structure of North American climate trends. A reason for the inability of the CMIP simulations to replicate key features of the observed spatial variations is revealed by diagnosing their simulated free atmospheric circulation trends, and comparing to the reanalysis data. The CMIP 500 mb height trends (Figure 3.7, middle panel) have little spatial structure, instead being dominated by a nearly uniform increase in heights. Given the strong thermodynamic relation between 500 mb heights and air temperature in the troposphere, the relative uniformity of North American surface warming in the CMIP simulations is consistent with the uniformity in its circulation change (there are additional factors that can influence surface temperature patterns, such as local soil moisture, snow cover and sea ice albedo [amount of short wave radiation reflected] effects on surface energy balances, that may have little influence in 500 mb heights).

In contrast, the ability of the AMIP simulations to produce key features of the observed spatial variations in surface temperature stems from the fact that SST variations during 1951 to 2006 force a trend in atmospheric circulation that projects upon the positive phases of both the PNA and NAO patterns (Figure 3.7, bottom panel). Although the amplitude of the ensemble-averaged AMIP 500 mb height trends is weaker than the observed 500 mb height trends, their spatial agreement is high. It is this spatially varying pattern of the the tropospheric circulation trend since 1951 that permits the reorganization of air mass movements and storm track shifts that is an important factor for explaining

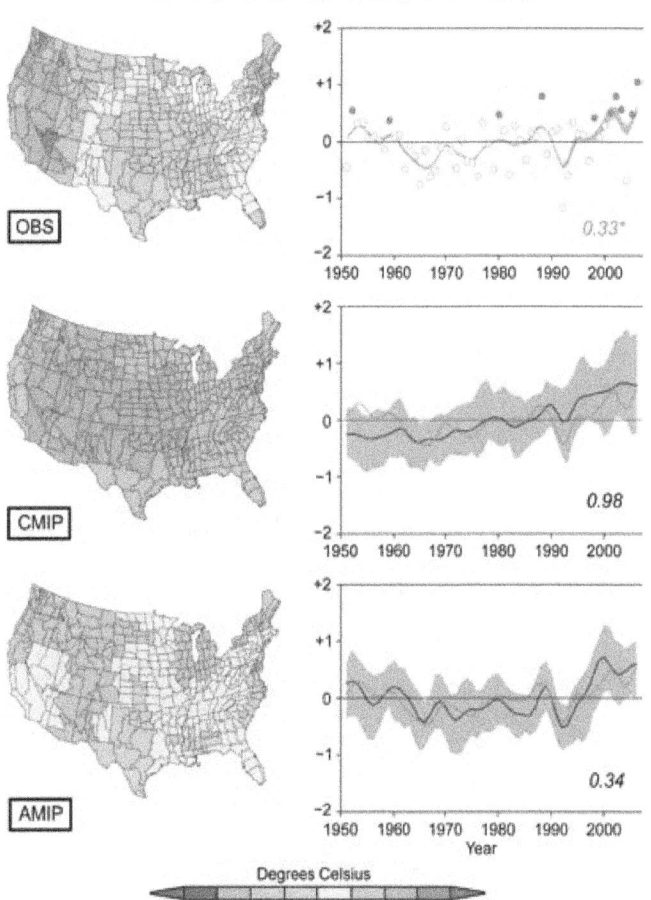

United States Summer Temperature: 1951-2006

Figure 3.8 Spatial maps of the linear temperature trend (°C total change over 56 year period) in summer (June-July-August) (left side) and time series of the variations over time of United States area-averaged temperatures in summer from observations, CMIP model simulations, and AMIP model simulations. Plotted values are the total 56-year change (°C), with the single asterisk denoting high confidence that an observed change was detected. Gray band in top panel denotes the range of observed temperatures based on five different analyses, gray band in middle panel denotes the 5 to 95 percent range among 41 CMIP model simulations, and gray band in lower panel denotes the 5 to 95 percent range among 33 AMIP model simulations. Curves are smoothed annual values using a five-point Gaussian filter. The Gaussian filter is a weighted time averaging applied to the raw annual values in order to highlight lower frequency variations. "Five-point" refers to the use of five annual values in the weighting process. Unsmoothed observed annual temperature anomalies are shown in open red circles, with warmest ten years shown in closed red circles.

key regional details of North American surface climate trends.

3.3.2.2 CONTIGUOUS UNITED STATES
For the U.S. area-averaged temperature variations, six of the warmest ten summers (Figure 3.8, top) and six of the warmest ten winters (Figure 3.9, top) during 1951 to 2006 occurred in the last decade (1997 to 2006). This recent cluster-

United States Winter Temperature: 1951-2006

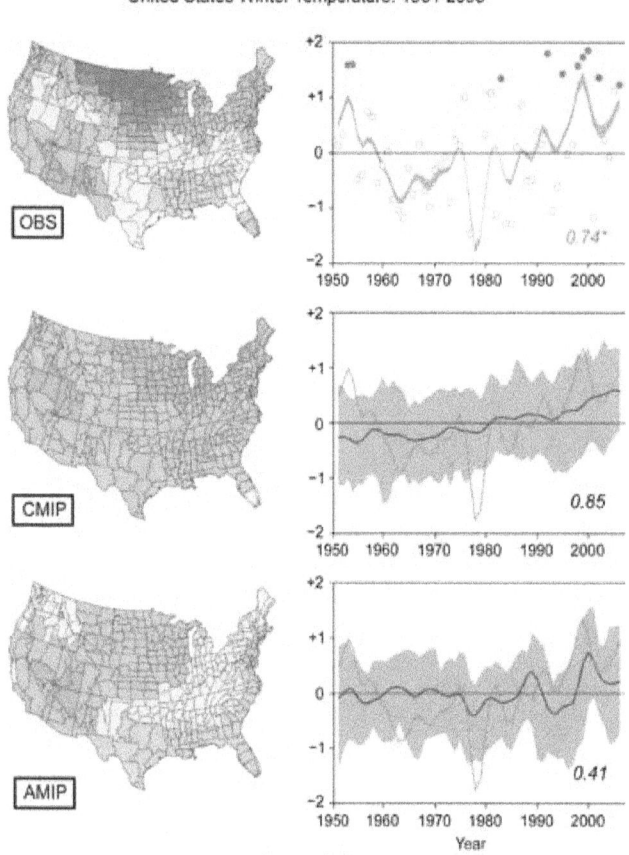

Figure 3.9 Spatial maps of the linear temperature trend (°C total over 56 year period) in winter (December-January-February) (left side) and time series of the variations over time of U.S. area-averaged temperatures in summer from observations, CMIP model simulations, and AMIP model simulations. Plotted values are the total 56-year change (°C), with the double asterisks denoting very high confidence that an observed change was detected. Gray band in top panel denotes the range of observed temperatures based on five different analyses, gray band in middle panel denotes the 5 to 95 percent range among 41 CMIP model simulations, and gray band in lower panel denotes the 5 to 95 percent range among 33 AMIP model simulations. Curves smoothed with five-point Gaussian filter. The Gaussian filter is a weighted time averaging applied to the raw annual values in order to highlight lower frequency variations. "Five-point" refers to the use of five annual values in the weighting process. Unsmoothed observed annual temperature anomalies are shown in open red circles, with warmest ten years shown in closed red circles.

ing of record warm occurrences is consistent with the increasing signal of anthropogenic greenhouse gas warming, as evidenced from the CMIP simulations (Figures 3.8 and 3.9, middle panels) that indicate accelerated warming over the United States during the past decade during both summer and winter.

During summer since 1951, some regions of the United States have observed strong warming while other regions experienced no significant change. The lack of mid-continent warming is a particularly striking feature of the observed trends since 1951, especially compared with the strong warming in the West. This overall pattern of U.S. temperature change is *unlikely* due to anthropogenic forcing alone, an assessment that is supported by several pieces of evidence. First, the spatial variations of the CMIP simulated U.S. temperature trend (Figure 3.8, middle) are not correlated with those observed—the pattern correlation is -0.10 (low and negative correlation) when removing the area-averaged warming. The ensemble CMIP area-averaged summer warming trend of +0.99°C (+1.78°F) is also three times higher than the observed area-averaged warming of +0.33°C (+0.59°F). In other words, there has been much less summertime warming observed for the United States as a whole than expected, based on changes in the external forcing. There is reason to believe, as discussed further below, that internal variations have been masking the anthropogenic greenhouse gas warming signal in summer to date, although the possibility that the simulated signal is too strong cannot be entirely ruled out.

Second, the spatial variations of the AMIP simulations for the U.S. temperature trend (Figure 3.8, bottom) are positively correlated with the observed observations, with a pattern correlation of +0.43 when the area-averaged warming is removed. The cooling of the southern Plains in the AMIP simulations agrees particularly well with observations. The reduced ensemble AMIP area-averaged U.S. summer warming trend of only +0.34°C (+0.61°F) is similar to observations. It thus appears that regional SST variability has played an important role in U.S. summer temperature trends since 1951. The nature of these important SST variations remains unknown. The extent to which they are due to internal coupled system variations and the contribution from anthropogenic forcing are among the vital questions awaiting future attribution research.

During winter, the pattern of observed surface temperature trends (Figure 3.9, top) consists of strong and significant warming over the western and northern United States, and insignificant change along the Gulf Coast in the South. Both CMIP and AMIP simulations produce key features of the U.S. temperature trend pattern (spatial correlations of 0.70 and 0.57, respectively, upon removing the U.S. area-averaged warming trend), although the cooling along the Gulf Coast appears inconsistent with external forcing, but consistent with SST forcing. The observed U.S. winter warming trend of +0.75°C (1.35°F) has been stronger than that occurring in summer, and compares to an area-averaged warming of +0.85°C (+1.53°F) in the ensemble of CMIP and +0.41°C (+0.74°F) in the ensemble of AMIP simulations.

It is worth noting that the United States also experienced warm conditions during the mid-twentieth century—the early years of available reanalyses (see also Box 3.3 for discussion of the warmth in the United States in the early twentieth century). This is an indication as to how sensitive trends are to the beginning and ending years selected for diagnosis, thus requiring that the trend analysis be accompanied by an assessment of the full evolution over time during 1951 to 2006.

Regarding confidence levels for the observed U.S. temperature trends for 1951 to 2006, a non-parametric test has been applied that estimates the probability distribution of 56-year trends attributable to natural variability alone (see Appendix B for methodological details). As in Section 3.2, this involves diagnosis of 56-year trends from the suite of "naturally forced" CMIP runs, from which a sample of 76 such trends were generated for the contiguous United States for winter and summer seasons. The observed area-averaged U.S. summer trend of +0.33°C (+0.59°F) is found to exceed the 80 percent level of trend occurrences in those natural forced runs, indicating a *high* level of confidence that warming has been detected. For winter, the observed trend of +0.75°C (+1.35°F) is found to exceed the 95 percent level of trends in the natural forced runs indicating a *very high* level of confidence. These diagnoses support this assessment that a warming of U.S. area-averaged temperatures during 1951 to 2006 has *likely* been detected for summer and *very likely* been detected for winter.

> A warming of U.S. area-averaged temperatures during 1951 to 2006 has *likely* been detected for summer and *very likely* been detected for winter.

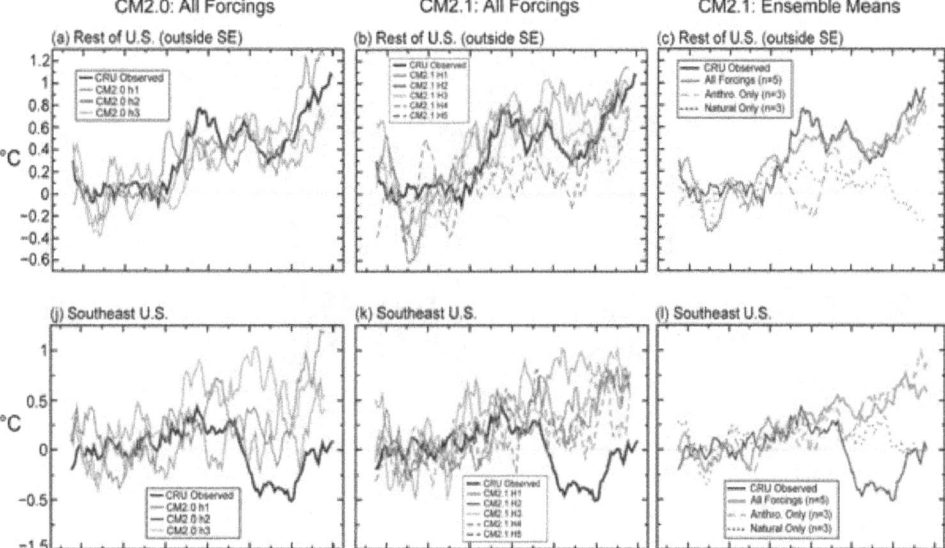

Figure 3.10 Ten-year running-mean area-averaged time series of surface temperature anomalies (°C) relative to 1881 to 1920 for observations and models for various regions: (a) through (c) rest of the contiguous United States, and (j) through (l) U.S. Southeast. The left column and middle columns are based on all-forcing historical runs 1871 to 2000 and observations 1871 to 2004 for GFDL coupled climate model CM2.0 (n=3) and CM2.1 (n=5), respectively. The right column is based on observed and model data through 2000, with ±2 standard error ranges (shading) obtained by sampling several model runs according to observed missing data. The red, blue, and green curves in the right-hand-column diagrams are ensemble mean results for the CM2.1 all-forcing (n=5), natural-only (n=3), and anthropogenic-only (n=3) forcing historical runs. Model data were masked according to observed data coverage. From Knutson *et al.* (2006).

Urbanization, land clearing, deforestation, and reforestation are likely to have contributed to some of the spatial patterns of warming over the United States.

The causes of the reduced warming in the U.S. Southeast compared to the remainder of the country, seen during both winter and summer seasons, have been considered in several studies. Knutson *et al.* (2006) contrasted the area-averaged temperature variations for the Southeast with variations for the remainder of the United States (as shown in Figure 3.10) for both observations and model simulations with the GFDL CM2 coupled model. While the observed and simulated warming due to anthropogenic forcing agrees well for the remainder of the United States, the observed cooling was outside the range of temperature variations that occurred among the small number of individual model simulations performed. For a larger ensemble size, such as provided by the whole CMIP multi-model ensemble as considered by Kunkel *et al.* (2006), the cooling in the Southeast is within the range of model simulated temperature variations but would have to be associated with a very large case of natural cooling superimposed on anthropogenic forced larger scale warming. Robinson *et al.*

(2002) and Kunkel *et al.* (2006) have shown that this regional cooling in the central and southeastern United States is associated with the model response to observed SST variations, particularly in the tropical Pacific and North Atlantic oceans, and is consistent with the additional assessment of AMIP simulations presented in this Section.

For the cold half of the year in particular, the Southeast cooling is also consistent with the trends in teleconnection patterns that were diagnosed from the reanalysis data.

Other studies have argued that land use and land cover changes are additional possible factors for explaining the observed spatial variations of warming over the United States since 1951. The marked increase of irrigation in the Central Valley of California and the northern Great Plains is likely to have lead to an increase (warming) in minimum temperatures and a reduced increase (lesser warming) in maximum temperatures in summer (Christy *et al.*, 2006;

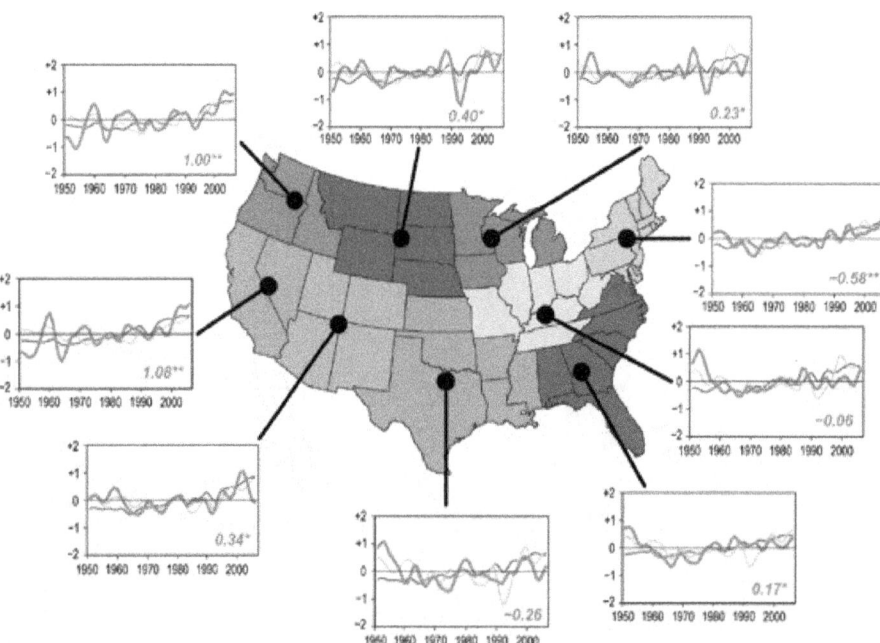

Figure 3.11 Regional U.S. surface temperature changes in summer (June-July-August) from 1951 to 2006. The observations are shown in bold red, ensemble-averaged CMIP in blue, and ensemble-averaged AMIP in green. A five-point Gaussian filter has been applied to the time series to emphasize multi-annual scale time variations. The Gaussian filter is a weighted time averaging applied to the raw annual values in order to highlight lower frequency variations. "Five-point" refers to the use of five annual values in the weighting process. Plotted values in each graph indicate the total 1951 to 2006 temperature change averaged for the sub-region. Double (single) asterisks denote regions where confidence of having detected a change is very high (high).

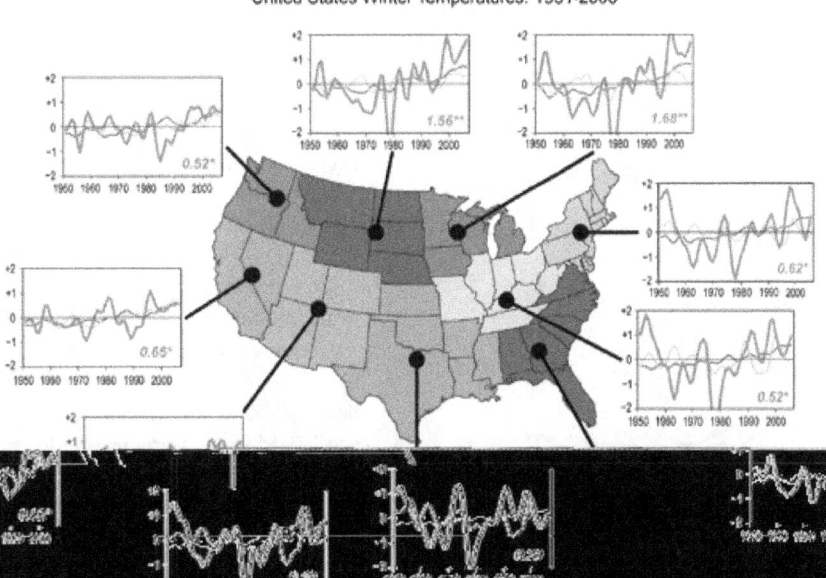

Figure 3.12 Regional U.S. surface temperature changes in winter (December-January-February) from 1951 to 2001. The observations are shown in bold red, ensemble-averaged CMIP in blue, and ensemble-averaged AMIP in green. A five-point Gaussian filter has been applied to the time series to emphasize multi-annual scale time variations. The Gaussian filter is a weighted time averaging applied to the raw annual values in order to highlight lower frequency variations. "Five-point" refers to the use of five annual values in the weighting process. Plotted values in each graph indicate the total 1951 to 2006 temperature change averaged for the sub-region. Single (double) asterisks denote regions where confidence of having detected a change is high (very high).

Kueppers *et al.*, 2007; Mahmood *et al.*, 2006). Urbanization, land clearing, deforestation, and reforestation are likely to have contributed to some of the spatial patterns of warming over the United States, though a quantification of these factors is lacking (Hale *et al.*, 2006; Kalnay and Cai, 2003; Trenberth, 2004; Vose *et al.*, 2004; Kalnay *et al.*, 2006).

As a further assessment of the spatial structure of temperature variations, the summer and winter surface temperature changes from 1951 to 2006 for nine U.S. subregions are shown in Figure 3.11 and 3.12, respectively. The observed temperature change is shown by the red bold curve, and the CMIP and AMIP ensemble-averaged temperature changes are given by blue and green curves, respectively. No attribution of recent climate variations and trends at these scales has been published, aside from the aforementioned Knutson *et al.* (2006) and Kunkel *et al.* (2006) studies that examined conditions over the U.S. Southeast. For decision making at these regional scales, as well as smaller local scales, a systematic explanation of such climate conditions is needed. In this

Product, several salient features of the observed and simulated changes are discussed; however, a complete synthesis has yet to be undertaken. For each region of the United States, the total 1951 to 2006 observed surface temperature change and its significance is plotted beneath the time series. Single asterisks denote high confidence and double asterisks denote very high confidence that a change has been detected using the methods described above.

During summer (Figure 3.11), there is *very high* confidence that warming has been observed over Pacific Northwest and Southwest regions. For these regions, the net warming since 1951 has been about +0.9°C (+1.6°F), exceeding the 95 percent level of trends in the natural forced runs at these regional levels. *High* confidence of a detected warming also exists for the Northeast, where the observed 56-year change is not as large, but occurs in a region of reduced variability, thereby increasing detectability of a change. These three warming regions also exhibit the best temporal agreement with the warming simulated in the CMIP models. In addition, the comparatively weaker observed

During summer, there is *very high* confidence that warming has been observed over Pacific Northwest and Southwest regions. For these regions, the net warming since 1951 has been about +0.9°C (+1.6°F).

summertime trends during 1951 to 2006 in the interior West, the southern Great Plains, the Ohio Valley, and the Southeast may be influenced by the very warm conditions at the beginning of the reanalysis record, a period

of widespread drought in those regions of the country.

During winter (Figure 3.12), there is *very high* confidence that warming has been detected over the northern Great Plains and the Great Lakes region. Confidence is *high* that warming during 1951 to 2006 has been detected in the remaining regions, except along the Gulf Coast in the South, where no detectable change in temperature has occurred. In the northern regions, most of the overall warming of about +1.5°C (+2.7°F) has happened in the last two decades. The CMIP simulations also produce accelerated winter warming over the northern United States in the past 20 years, suggesting that this regional and seasonal feature may have been influenced by anthropogenic forcing.

The 1950s produced some of the warmest winters during the 1951 to 2006 period for several regions of the U.S. The latest decade of warmth in the four southern and eastern United States regions still fails to exceed that earlier decadal warmth. The source for the warm winters in those regions in mid-century is not currently known, and it is unclear whether it is related to a widespread warm period across the Northern Hemisphere during the 1930s and 1940s that was attributed primarily to internal variability (Delworth and Knutson, 2000). The fact that neither CMIP nor AMIP ensemble-averaged responses produce 1950s warmth supports an interpretation that this warmth was likely un-related to external or the SST forcing.

3.3.3 Precipitation Trends
3.3.3.1 NORTH AMERICA
The observed annual North American precipitation trends during 1951 to 2006 in Figure 3.6 are dominated by regional scale features. The prominent identifiable features of change are the annual drying of Mexico and the greater Caribbean region, and the increase over northern Canada. However, due to the strong and differing seasonal cycles of precipitation across the continent, a diagnosis of the annually averaged trends is of limited value. Therefore, this Section focuses further discussion on the seasonal and regional analyses.

North American Winter Circulation and Precipitation Change

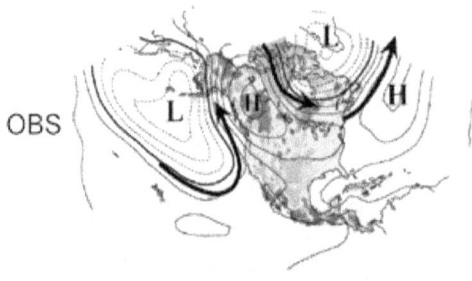
OBS

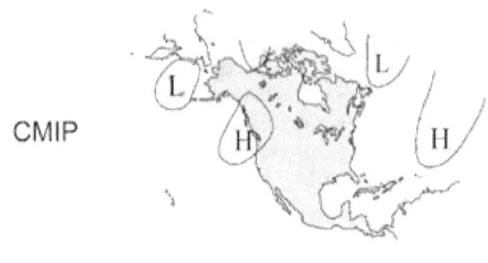
CMIP

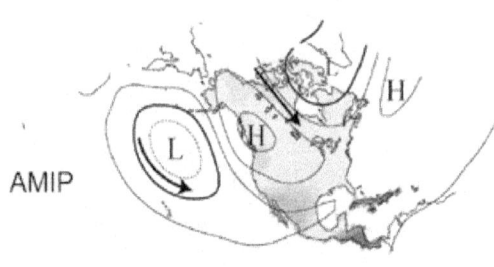
AMIP

Percent of Climatalogy

-50 -30 -25 -20 -15 -10 0 10 15 20 25 30 50

Figure 3.13 The 1951 to 2006 November to April trend of 500 millibar heights (contours, units meters total over 56 year period, contour interval 10 meters) and North American precipitation (color shading, units 56-year change as percent of the 1951 to 2006 climatological average) for observations (top), CMIP ensemble-averaged (middle), AMIP ensemble-averaged (bottom). Anomalous High and Low Pressure regions are highlighted. Arrows indicate the anomalous wind direction, which circulates around the High (Low) Pressure centers in a clockwise (counterclockwise) direction.

The cold-season (November to April) North American observed precipitation change is shown in Figure 3.13 (top), with superimposed contours of the tropospheric circulation change (identical to Figure 3.7). The reanalysis data of circulation change provides physical insights on the origins of the observed regional precipitation change. The band of drying that extends from British Columbia across much of southern Canada and part of the northern United States corresponds to upper level high pressure from which one can infer reduced storminess. In contrast, increased precipitation across the southern United States and northern Mexico in winter is consistent with the deeper southeastward shifted Aleutian low, a semi-permanent low pressure system situated over the Aleutian Islands in winter, that is conducive for increased winter storminess across the southern region of the United States. Further south, drying again appears across southern Mexico and Central America. This regional pattern is unrelated to external forcing alone, as revealed by the lack of spatial agreement with the CMIP trend pattern (middle panel), and the lack of a wavy tropospheric circulation response in the CMIP simulations. However, many key features of the observed regional precipitation change are consistent with the forced response to global SST variations during 1951 to 2006, as is evident from the AMIP trend pattern (bottom). In particular, the AMIP simulations generate the zonal band of enhanced high latitude precipitation, the band of reduced precipitation centered along 45°N, wetness in the southern United States and northern Mexico, and dryness over Central America. These appear to be consistent with the SST forced change in tropospheric circulation. Thus, in future attribution research it is important to determine the responsible regional SST variations, and to assess the origin of the SSTs anomalies themselves.

3.3.3.2 CONTIGUOUS UNITED STATES
The observed seasonally-averaged precipitation trends over the period 1951 to 2006 are compared with the ensemble-averaged responses of the CMIP and AMIP simulations for summer in Figure 3.14 and for winter in Figure 3.15. In general, during all seasons there are smaller scale spatial variations of the observed precipitation trends across the United States than for the temperature trends, and larger interannual

United States Summer Precipitation: 1951-2006

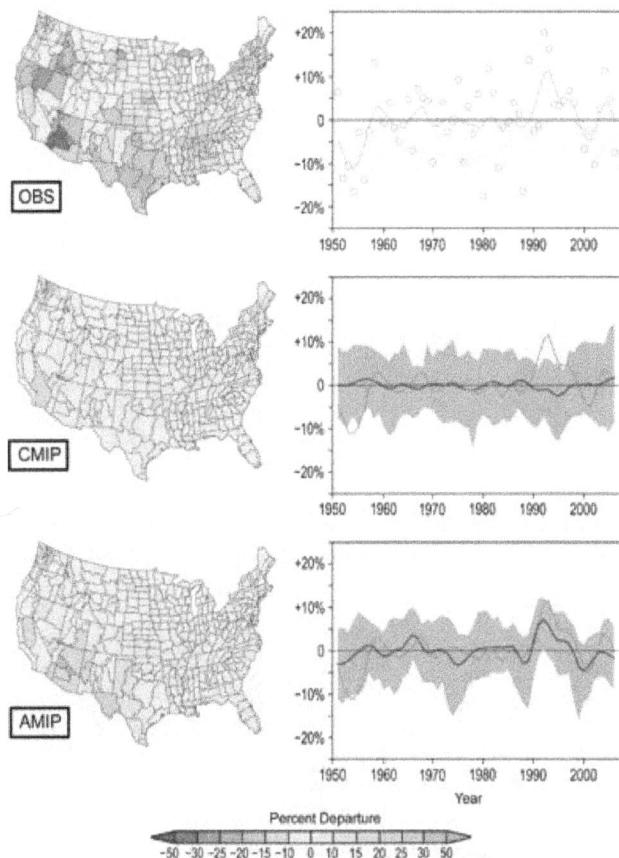

Figure 3.14 Spatial maps of the linear trend in precipitation (percent change of seasonally averaged 1951 to 2006 climatology) in summer (June-July-August) (left side) and the variations over time of U.S. area-averaged precipitation in summer from observations, CMIP model simulations, and AMIP model simulations. Gray band in middle panel denotes the 5 to 95 percent range among 41 CMIP model simulations, and gray band in lower panel denotes the 5 to 95 percent range among 33 AMIP model simulations. Curves smoothed using a five-point Gaussian filter. The Gaussian filter is a weighted time averaging applied to the raw annual values in order to highlight lower frequency variations. "Five-point" refers to the use of five annual values in the weighting process. Unsmoothed observed annual precipitation anomalies are shown in open red circles.

and decadal variability. These factors undermine the detectability of any physical change in precipitation since 1951.

During summer (Figure 3.14), there is a general pattern of observed rainfall reductions in the U.S. West and Southwest and increases in the East. There is some indication of similar patterns in the CMIP and AMIP simulations, however, the amplitudes are so weak that the ensemble model anomalies are themselves unlikely to be significant. The time series of U.S. summer rainfall is most striking for a recent

United States Winter Precipitation: 1951-2006

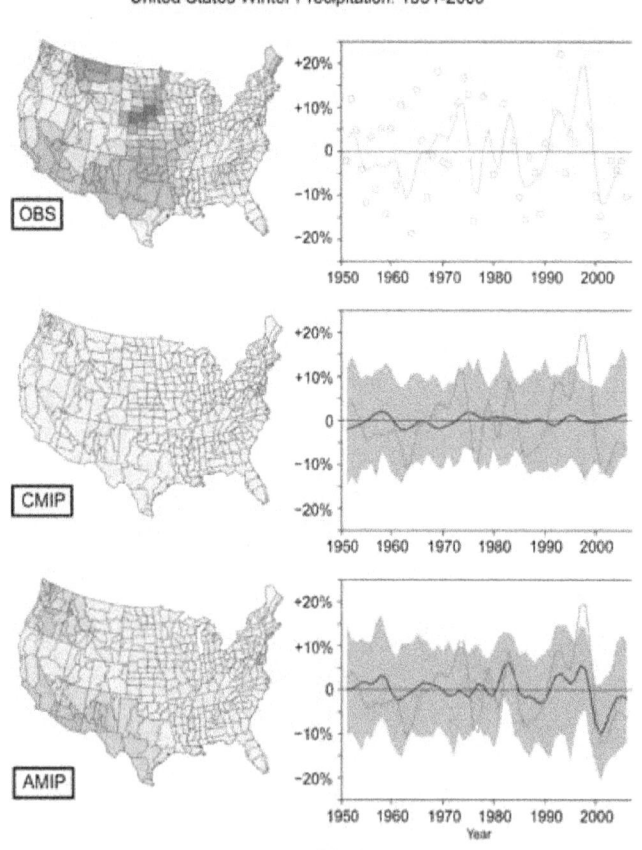

Figure 3.15 Spatial maps of the linear trend in precipitation (percent change of seasonal climatology) in winter (December-January-February) (left side) and the variations over time of U.S. area-averaged precipitation in winter from observations, CMIP model simulations, and AMIP model simulations. Gray band in middle panel denotes the 5 to 95 percent range among 41 CMIP model simulations, and gray band in lower panel denotes the 5 to 95 percent range among 33 AMIP model simulations. Curves smoothed using a five-point Gaussian filter. The Gaussian filter is a weighted time averaging applied to the raw annual values in order to highlight lower frequency variations. "Five-point" refers to the use of five annual values in the weighting process. Unsmoothed observed annual precipitation anomalies are shown in open red circles.

During winter (Figure 3.15), there is little agreement between the observed and CMIP modeled spatial patterns of trends, though considerably better agreement exists with the AMIP modeled spatial pattern. Again, the ensemble-averaged CMIP model simulations shows no significant long term trends during 1951 to 2006, and they also exhibit weak variability (middle), suggesting that changes in external forcing have had no appreciable influence on area-averaged precipitation in the United States. This is consistent with the published results of Zhang *et al.* (2007) who find disagreement between observed and CMIP simulated trends over the United States. In contrast, several key decadal variations are captured by the ensemble mean AMIP simulations including again the swing from wet 1990s to dry late 1990s early 2000 conditions. For the 56-year period as a whole, the temporal correlation of AMIP simulated and observed winter U.S. average rainfall is +0.59.

For the nine separate U.S. regions, Figures 3.16 and 3.17 illustrate the variations over time of observed, ensemble CMIP, and ensemble AMIP precipitation for summer and winter seasons, respectively. These highlight the strong temporal swings in observed regional precipitation between wet and dry periods, such that no single region has a detectable change in precipitation during 1951 to 2006. These observed fluctuations are nonetheless of great societal relevance, being associated with floods and droughts having catastrophic local impacts. Yet, comparing to CMIP simulations indicates that it is *exceptionally unlikely* that these events are related to external forcing. There is some indication from the AMIP simulations that their occurrence is somewhat determined by SST events, especially in the South and West, during winter presumably related to the ENSO cycle.

Other statistical properties of rainfall, including extremes in daily amounts and the fraction of annual rainfall due to individual wet days have exhibited a detectable change over the United States in recent decades, and such changes have been attributed to anthropogenic forcing in the companion CCSP SAP 3.3 Product (CCSP, 2008).

fluctuation between wet conditions in the 1990s, followed by dry conditions in the late 1990s and early 2000s. This prominent variation is well explained by the region's summertime response to SST variations, as seen by the remarkable correspondence of observations with the time evolving AMIP rainfall (lower panel). For the 56-year period as a whole, the temporal correlation of AMIP simulated and observed summer U.S. average rainfall is +0.64.

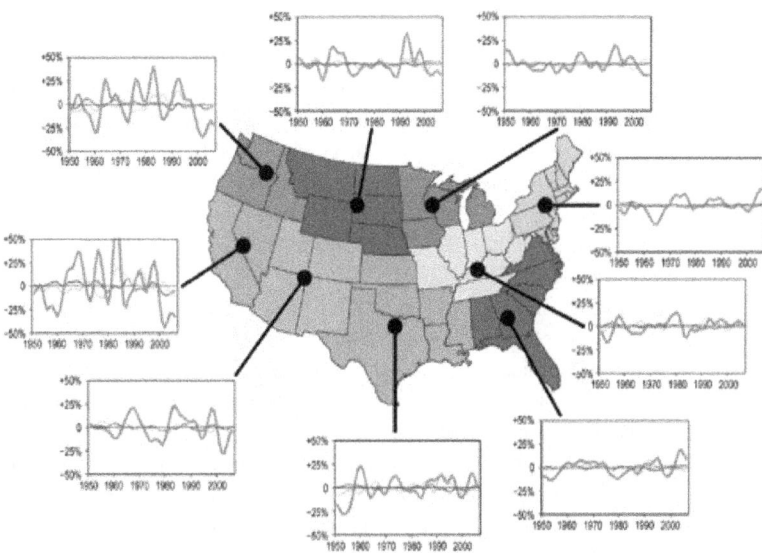

Figure 3.16 The 1951 to 2006 regional U.S. precipitation changes over time in summer (June-July-August). The observations are shown in bold red, ensemble-averaged CMIP in blue, and ensemble-averaged AMIP in green. A five-point Gaussian filter has been applied to the time series to emphasize multi-annual scale time variations. The Gaussian filter is a weighted time averaging applied to the raw annual values in order to highlight lower frequency variations. "Five-point" refers to the use of five annual values in the weighting process.

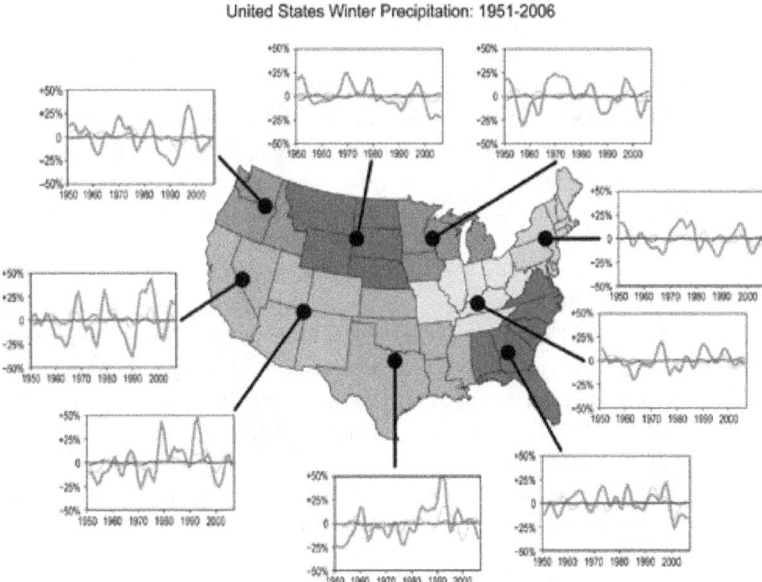

Figure 3.17 The 1951 to 2006 regional U.S. precipitation changes over time in winter (December-January-February). The observations are shown in bold red, ensemble-averaged CMIP in blue, and ensemble-averaged AMIP in green. A five-point Gaussian filter has been applied to the time series to emphasize multi-annual scale time variations. The Gaussian filter is a weighted time averaging applied to the raw annual values in order to highlight lower frequency variations. "Five-point" refers to the use of five annual values in the weighting process.

3.4 NATURE AND CAUSE OF APPARENT RAPID CLIMATE SHIFTS FROM 1951 TO 2006

3.4.1 Introduction
Rapid climate shifts are of scientific interest and of public concern because of the expectation that such occurrences may be particularly effective in exposing the vulnerabilities of societies and ecosystems (Smith *et al.*, 2001). Such abrupt shifts are typically distinguished from the gradual pace of climate change associated, for instance, with anthropogenic greenhouse gas forcing. However, through non-linear feedbacks, gradual forcing could also trigger rapid shifts in some parts of the climate system, a frequently cited example being a possible collapse of the global ocean's principal conveyor of heat between the tropics and high latitudes known as the thermohaline circulation (Clarke *et al.*, 2002).

By their very nature, abrupt shifts are unexpected events—climate surprises—and thus offer particular challenges to policy makers in planning for their impacts. A retrospective assessment of such "rare" events may offer insights on mitigation strategies that are consistent with the severity of impacts related to rapid climate shifts. Such an assessment would also consider impacts of abrupt climate shifts on societies and ecosystems and would also prepare decision makers to anticipate consequences of gradual changes in climate, insofar as they may be no less severe than those related to rapid climate shifts.

3.4.2 Defining Rapid Climate Shifts
A precise definition for a climate shift that is either "rapid" or "abrupt" does not exist because there is limited knowledge about the full sensitivity of the climate system. For instance, due to nonlinearity, changes in external forcing may not lead to a proportionate climate response. It is conceivable that a *gradual* change in external forcing could yield an abrupt response when applied near a tipping point (the point at which a slow gradual change becomes irreversible and then proceeds at a faster rate of change) of sensitivity in the climate system, whereas an *abrupt* change in forcing may not lead to any abrupt response when it is applied far from the system's tipping point. To date, little is known

A rapid climate shift is one occurring so fast that societies and ecosystems have difficulty adapting to it.

about the threshold tipping points of the climate system (Alley *et al.*, 2003).

In its broadest sense, a "rapid" shift is a transition between two climatic states that individually have much longer duration than the transition period itself. From an impacts viewpoint, a rapid climate shift is one occurring so fast that societies and ecosystems have difficulty adapting to it.

3.4.3 Mechanisms for Rapid Climate Shifts
The National Research Council (NRC, 2002) has undertaken a comprehensive assessment of rapid climate change, summarizing evidence of such changes occurring before the instrumental and reanalysis records, and understanding abrupt changes in the modern era. The NRC (2002) report on abrupt climate change draws attention to evidence for severe swings in climate proxies of temperature (so-called paleoreconstructions) during both the last ice age and the subsequent interglacial period known as the Holocene. Ice core data indicate that abrupt shifts in climate have often occurred during Earth's climate history, indicating that gradual and smooth movements do not always characterize climate variations. Identification of such shifts is usually empirical, based upon expert assessment of long time series of the relevant climate records, and in this regard, their recognition is retrospective. Against this background of abundant evidence for the magnitude of rapid climate shifts, there is a lack of information about the mechanisms that can lead to climate shifts and of the processes by which climate is maintained in various altered states (Broecker, 2003). Understanding the causes of such shifts is a prerequisite to any early warning system that is, among other purposes, needed for planning the scope and pace of mitigation.

The National Academy report (NRC, 2002) also highlights three possible mechanisms for abrupt change: (1) an abrupt forcing, such as may occur through meteorite impacts or volcanic eruptions; (2) a threshold-like sensitivity of the climate system in which sudden changes can occur even when subjected to gradual changes in forcing; and (3) an unforced behavior of the climate system resulting purely from chaotic internal variations.

3.4.4 Rapid Climate Shifts since 1950

Although changes in external forcing, whether natural or anthropogenic, are not yet directly assimilated in the current generation of reanalysis products, abrupt changes in external forcings can still influence the reanalyses indirectly through their effect on other assimilated variables. Observational analyses of the recent instrumental record give some clues of sudden climate shifts, characterized as those that have had known societal consequences. These are summarized below according to the current understanding of the potential mechanism involved. For several reasons, the sustainability of these apparent shifts is not entirely known. First, since 1950, multi-decadal fluctuations are readily seen in North American temperatures (Figure 3.3) and precipitation (Figure 3.6). Although the post-1950 period is the most accurately observed period of Earth's climate history, the semi-permanency of any change cannot be readily judged from merely 50 years of data. This limited perspective of our brief modern climate record stands in contrast to proxy climate records, within which stable climate was punctuated by abrupt change leading to new climate states lasting centuries to millennia. Second, it is not known whether any recent rapid transitions have involved threshold exceedences in a manner that would forewarn of their permanence.

3.4.4.1 ABRUPT NATURAL EXTERNAL FORCINGS SINCE 1950

The period of the reanalysis record was a volcanically active one, particularly compared with the first half of the twentieth century. Three major volcanic eruptions included the Agung in 1963, El Chichon in 1982, and Mt. Pinatubo in 1991. Each eruption injected aerosols into the stratosphere (about 10 kilometers, or 6 miles, above the Earth's surface), acting to significantly increase the stratospheric aerosol optical depth that led to an increase in the reflectance of incoming solar radiation (Santer *et al.*, 2006).

Each of these abrupt volcanic forcings has been found to exert a discernable impact on climate conditions. Observed sea surface temperatures cooled in the wake of the eruptions, the detect-

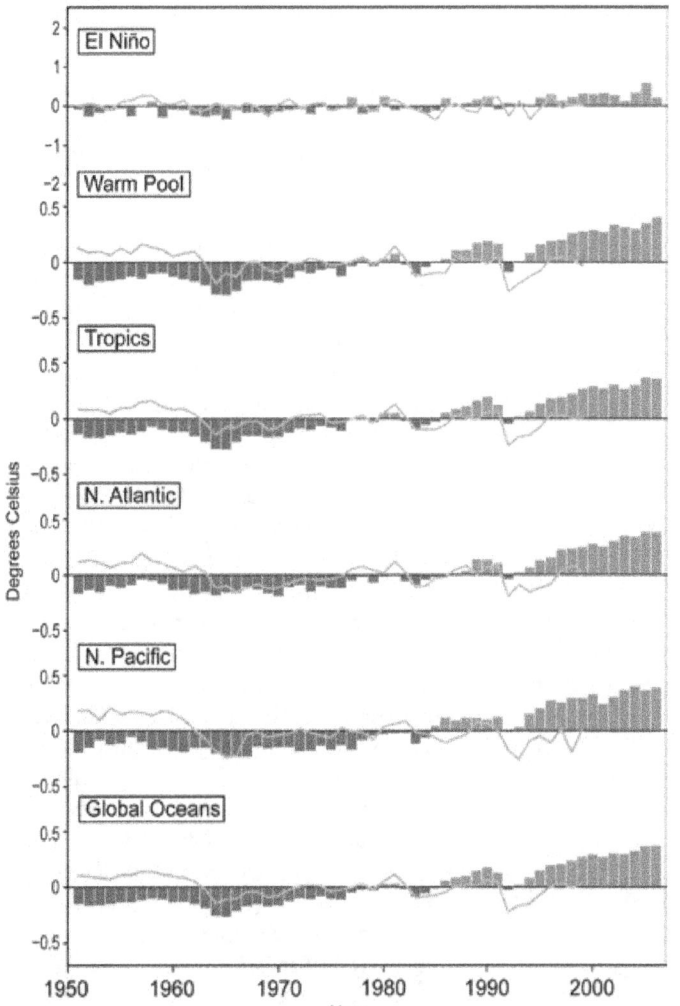

CMIP Annual SST Time Series

Figure 3.18 CMIP simulated annually-averaged SST changes over time for 1951 to 2006. The oceanic regions used to compute the indices are 5°N to 5°S, 90°W to 150°W for El Niño, 10°S to 10°N, 60°E to 150°E for the warm pool, 30°S to 30°N for the tropics, 30°N to 60°N for the North Atlantic, 30°N to 60°N for the North Pacific, and 40°S to 60°N for the global oceans. Dataset is the ensemble average of 19 CMIP models subjected to the combination of external anthropogenic and natural forcing, and anomalies are calculated relative to each model's 1951 to 2006 reference. Green curve is the surface temperature change based on the ensemble average of four CMIP models forced only by time evolving natural forcing (volcanic and solar).

ability of which was largest in oceans having small unforced, internal variability (Santer *et al.*, 2006). Surface-based observational analyses of these and other historical volcanoes indicate that North American surface temperatures tend to experience warming in the winters following strong eruptions, but cooling in the subsequent summer (Kirchner *et al.*, 1999). However, these abrupt forcings have not led to sustained changes in climate conditions, namely because the residence time for the stratospheric aerosol increases due to volcano eruption is less than a few years (depending on the particle sizes and the geographical location of the volcanic eruption), and the fact that major volcanic events since 1950 have been well separated in time.

The impact of the volcanic events is readily seen in Figure 3.18 (green curve) which plots annual SST changes over time in various ocean basins derived from the ensemble-averaged CMIP simulations forced externally by estimates of the time evolving volcanic and solar forcings (so-called "natural forcing" runs). The SST cooling in the wake of each event is evident. Furthermore, in the comparison with SST evolutions in the fully forced natural and anthropogenic CMIP runs (Figure 3.18, bars), the lull in ocean warming in the early 1980s and early 1990s was likely the result of the volcanic aerosol effects. Similar lulls in warming rates are evident in the observed SSTs at these times (Figure 3.5). They are also evident in the observed and CMIP simulated North American surface temperature changes over time (Figure 3.3). Yet, while having detected the climate system's response to abrupt forcing, and while some model simulations detect decade-long reductions in oceanic heat content following volcanic eruptions (Church *et al.*, 2005), their impacts on surface temperature have been relatively brief and transitory.

3.4.4.2 ABRUPTNESS RELATED TO GRADUAL INCREASE OF GREENHOUSE GASES SINCE 1950

Has the gradual increase in greenhouse gas external forcing triggered threshold-like behavior in climate, and what has been the relevance for North America? There is evidence of abrupt changes of ecosystems in response to anthropogenic forcing that is consistent with tipping point behavior over North America (Adger *et al.*, 2007). Some elements of the physical climate system including sea ice, snow cover, mountainous snowpack, and streamflow have also exhibited rapid change in recent decades (IPCC, 2007a).

There is also some suggestion of abrupt change in ocean surface temperatures. Whereas the overall global radiative forcing due to increasing greenhouse gases has increased steadily since

1950 (IPCC, 2007a), observed sea surface temperature over the warmest regions of the world ocean—the so-called warm pool—experienced a rapid shift to warm conditions in the late 1970s (Figure 3.5). In this region covering the tropical Indian Ocean/West Pacific where surface temperatures can exceed 30°C (86°F), the noise of internal SST variability is weak, increasing the confidence in the detection of change. While there is some temporal correspondence between the rapid 1970s emergent warm pool warming in observations and CMIP simulations (Figure 3.18), further research is required to confirm that a threshold-like response of the ocean surface heat balance to steady anthropogenic forcing occurred.

The matter of the relevance of abrupt oceanic warming for North American climate is even less clear. On one hand, North American surface temperatures also warmed primarily after the 1970s, although not in an abrupt manner. The fact that the AMIP simulations yield a similar behavior suggests some cause-effect link to the oceans. On the other hand, the CMIP simulations generate a steadier rate of North American warming during the reanalysis period, punctuated by brief pauses due to volcanic aerosol-induced cooling events.

3.4.4.3 ABRUPTNESS DUE TO UNFORCED CHAOTIC BEHAVIOR SINCE 1950

Some rapid climate transitions in recent decades appear attributable to chaotic natural fluctuations. One focus of studies has been the consequence of an apparent shift in the character of ENSO events after the 1970s, with more frequent El Niño warming in recent decades (Trenberth and Hoar, 1996).

Abrupt decreases in rainfall occurred over the U.S. Southwest and Mexico in the 1950s and 1960s (Narisma *et al.*, 2007), with a period of enhanced La Niña conditions during that decade being a likely cause (Schubert *et al.*, 2004; Seager *et al.*, 2005). However, this dry period, and the decadal period of the Dust Bowl that preceded it over the Great Plains, did not constitute permanent declines in those regions' rainfall, despite meeting some criteria for detecting abrupt rainfall changes (Narisma *et al.*, 2007). In part, the ocean conditions that

Some rapid climate transitions in recent decades appear attributable to chaotic natural fluctuations.

contributed to these droughts did not persist in their cold La Niña state.

An apparent rapid transition of the atmosphere-ocean system over the North Pacific was observed to occur in the period from 1976 to 1977. From an oceanographic perspective, changes in ocean heat content and SSTs that happened suddenly over the Pacific basin north of 30°N were caused by atmospheric circulation anomalies (Miller *et al.*, 1994). These consisted of an unusually strong Aleutian Low that developed in the fall season of 1976, a feature that recurred during many successive winters for the next decade (Trenberth, 1990). These surface features were linked with a persistent positive phase of the PNA teleconnection pattern in the free atmosphere as revealed by reanalysis data. The time series of wintertime Alaskan surface temperatures (Figure 3.19) reveals the mild conditions that suddenly emerged after 1976. This transition in climate was accompanied by significant shifts in marine ecosystems throughout the Pacific Basin (Mantua *et al.*, 1997). It is now evident that this Pacific Basin-North American event, while perhaps meeting some criteria for a rapid transition, was mostly due to a large scale coupled-ocean atmosphere variation over multiple decades (Latif and Barnett, 1996). Thus, it is best viewed as a climate "variation" rather than as an abrupt change in the coupled ocean-atmosphere system (Miller *et al.*, 1994). Such multidecadal variations are readily seen in the observed index of both the North Pacific and the North Atlantic SSTs. However, the Alaskan temperature time series also indicates that there has been no return to cooler surface conditions in recent years. While the pace of anthropogenic warming alone during the last half-century has been more gradual than the rapid warming observed over Alaska, the superposition of an internal decadal fluctuation can lend the appearance of an abrupt warming, as Figure 3.19 indicates occurred over western North America in the mid-1970s. It is plausible that the permanency of the shifted surface warmth is rendered by the progressive increase in the strength of the external anthropogenic signal relative to the amplitude of internal decadal variability.

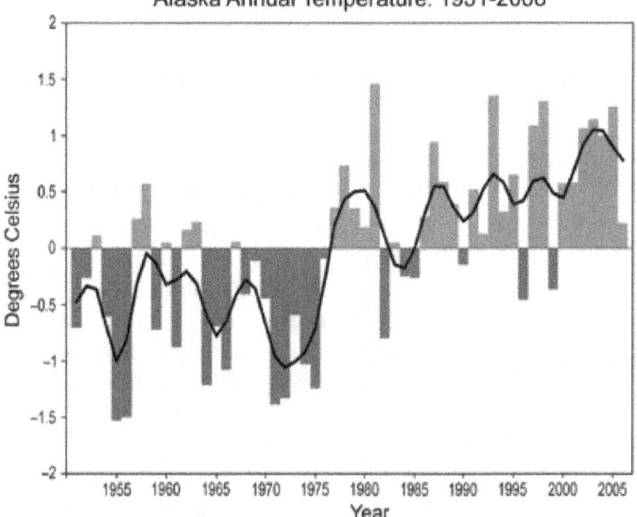

Figure 3.19 Observed Alaska annual surface temperature departures for 1951 to 2006. Anomalies are calculated relative to a 1951 to 2006 reference. Smoothed curve is a five-point Gaussian filter applied to the annual departures to emphasize multi-annual variations. The Gaussian filter is a weighted time averaging applied to the raw annual values in order to highlight lower frequency variations. "Five-point" refers to the use of five annual values in the weighting process.

3.5 UNDERSTANDING OF THE CAUSES FOR NORTH AMERICAN HIGH-IMPACT DROUGHT EVENTS FOR 1951 TO 2006

3.5.1. Introduction
Climate science has made considerable progress in understanding the processes leading to drought, due in large part to the emergence of global observing systems. The analysis of the observational data reveals relationships with large-scale atmospheric circulation patterns, and illustrates linkages with sea surface temperature patterns as far away from North America as the equatorial Pacific and Indian Ocean. Computing capabilities to perform extensive experimentation—only recently available—are permitting first ever quantifications of the sensitivity of North American climate to various forcings, including ocean temperatures and atmospheric chemical composition.

Such progress, together with the recognition that the U.S. economy suffers during severe droughts, has led to the launch of a National Integrated Drought Information System (NIDIS, 2004), whose ultimate purpose is to develop a timely and useful early warning system for drought.

Credible prediction systems are always improved when supported by knowledge of the underlying mechanisms and causes for the phenomenon's variability.

The North American continent has experienced numerous periods of drought during the reanalysis period, 1951 to 2006.

Credible prediction systems are always improved when supported by knowledge of the underlying mechanisms and causes for the phenomenon's variability. In this Section, current understanding of the origins of North American drought is assessed, focusing on events during the period of abundant global observations since about 1950. Assessments of earlier known droughts (such as the Dust Bowl) serve to identify potential cause-effect relationships that may apply to more recent and future North American regional droughts, and this perspective is provided as well (see Box 3.3 for discussion of the Dust Bowl).

3.5.2 Definition of Drought

Many definitions for drought appear in the literature, each reflecting its own unique social and economic context in which drought information is desired. In this Product, the focus is on meteorological drought as opposed to the numerous impacts (and measures) that could be used to characterize drought (*e.g.*, the hydrologic drought, indicated by low river flow and reservoir storage, or the agricultural drought, indicated by low soil moisture and deficient plant yield).

Meteorological drought has been defined as "a period of abnormally dry weather sufficiently prolonged for the lack of water to cause serious hydrologic imbalance in the affected area" (Huschke, 1959). The policy statement of the American Meteorological Society defines meteorological and climatological drought in terms of the magnitude of a precipitation

shortfall and the duration of this shortfall event (AMS, 2004).

The Palmer Drought Severity Index (PDSI) (Palmer, 1965) measures the deficit in moisture supply relative to its demand at the Earth's surface, and is used in this Chapter to illustrate some of the major temporal variations of drought witnessed over North America. The Palmer Drought Index is also useful when intercomparing historical droughts over different geographical regions (*e.g.*, Karl, 1983; Diaz, 1983), and it has been found to be a useful proxy of soil moisture and streamflow deficits that relate to the drought impacts having decision-making relevance (*e.g.*, Dai *et al.*, 2004).

3.5.3 Drought Causes
3.5.3.1 DROUGHT STATISTICS, MECHANISMS AND PROCESSES

The North American continent has experienced numerous periods of drought during the reanalysis period. Figure 3.20 illustrates the time variability of areal coverage of severe drought since 1951, and on average, 10 percent (14 percent) of the area of the contiguous (western) United States experiences severe drought each year. The average PDSI for the western states during this time period is shown in the bottom panel; while it is very likely dominated by internal variability, the severity of the recent drought compared with others since 1950 is also apparent.

The middle of the twentieth century began with severe drought that covered much of the United

BOX 3.5: Drought Attribution and Use of Reanalysis Data

The indications for drought itself, such as the Palmer Drought Severity Index (PDSI) or precipitation, are not derived from reanalysis data, but from the network of surface observations. The strength of reanalysis data lies in its depiction of the primary variables of the free atmospheric circulation and linking them with the variability in the PDSI. As discussed in Chapter 3, the development and maintenance of atmospheric ridges is the prime ingredient for drought conditions, and reanalysis data is useful for understanding the etymology of such events: their relationship to initial atmospheric conditions, potential downstream and upstream linkages, and the circulation response to soil moisture deficits and SST anomalies. Many drought studies compare model simulations of hypothetical causes to observed atmospheric circulation parameters; reanalysis data can help differentiate among the different possible causes by depicting key physical processes by which drought events evolved.

For final attribution, the drought mechanism must be related to either a specific forcing or internal variability. Reanalysis data, available only since about 1950, is of too short a length to provide a firm indication of internal variability. It also does not indicate (or utilize) direct impact of changing climate forcings, such as increased greenhouse gases or varying solar irradiance. The relationship of atmospheric circulation changes to these forcings must be provided by empirical correlation or, better yet, General Circulation Model (GCM) studies where cause and effect can be directly related.

Conterminous U.S. Drought Coverage

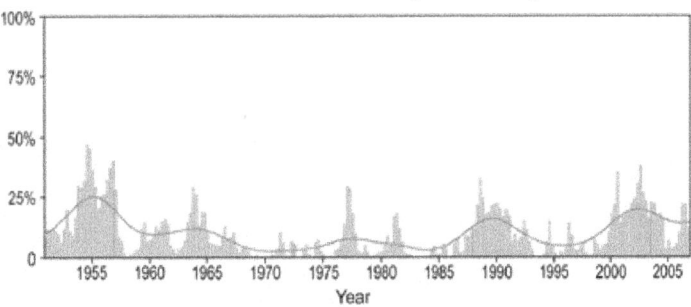

Western U.S. Drought Coverage

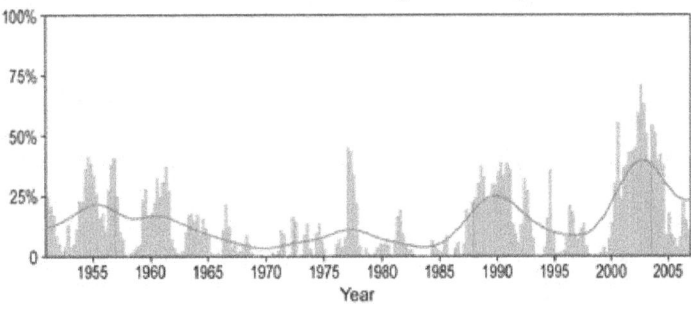

Western U.S. Average PDSI

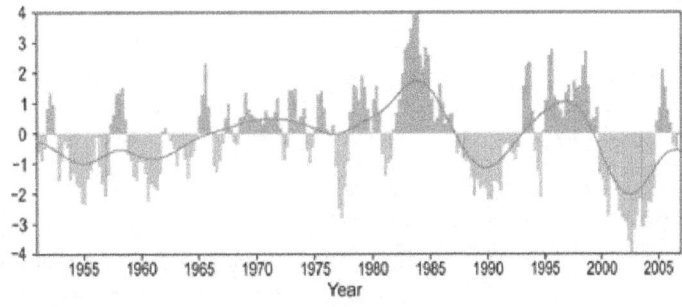

Figure 3.20 Percentage of contiguous United States (top) and western United States (middle) covered by severe or extreme drought, as defined by Palmer Drought Severity Index (PDSI) as less than -3. Time series of the western United States area-averaged PDSI. Positive (Negative) PDSI indicative of above (below) average surface moisture conditions. The western United States consists of the 11 western-most contiguous U.S. states. Red lines depict the time series smoothed using a nine-point Gaussian filter in order to emphasize lower frequency variations. The Gaussian filter is a weighted time averaging applied to the raw annual values. "Five-point" refers to the use of five annual values in the weighting process.

States. Figure 3.21 illustrates the observed surface temperature (top) and precipitation anomalies (bottom) during the early 1950s drought. The superimposed contours are of the 500 mb height from reanalysis data that indicates one of the primary causal mechanisms for drought: high pressure over and upstream that steers moisture-bearing storms away from the drought-affected region.

The northeastern United States had severe drought from about 1962 to 1966, with dry conditions extending southwestward into Texas. The 1970s were relatively free from severe drought, and since 1980 there has been an increased frequency of what the National Climatic Data Center (NCDC) refers to as "billion dollar United States weather disasters", including several major drought events: (1) Summer 1980, central/eastern United States; (2) Summer 1986,

1951-1956 Annual Composite

Temperature

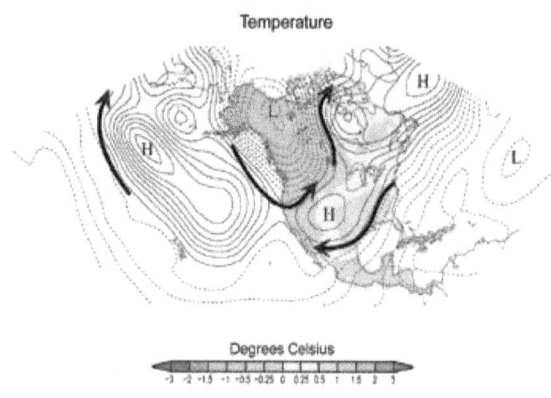

Degrees Celsius

Precipitation

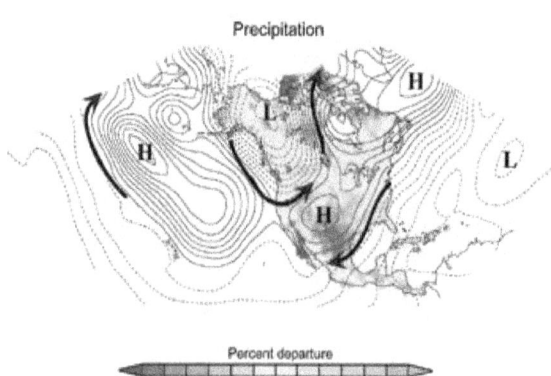

Percent departure

Figure 3.21 Observed climate conditions averaged for 1951 to 1956 during a period of severe U.S. Southwest drought. The 500 millibar height field (contours, units\meters) is from the NCEP/NCAR R1 reanalysis. The shading indicates the five-year average anomaly of the surface temperature (top) and precipitation (bottom). The surface temperature and precipitation are from independent observational datasets. Anomalous High and Low Pressure regions are highlighted. Arrows indicate the anomalous wind direction, which circulates around the High (Low) Pressure centers in a clockwise (counterclockwise) direction.

ability occurs along the 95°W meridian, while the lowest variability relative to the average precipitation is in the northeast, a distribution that parallels the occurrence of summertime droughts. This picture is somewhat less representative of droughts in the western United States, a region which receives most of its precipitation during winter.

It is natural to ask whether the plethora of recent severe drought conditions identified by NCDC is associated with human effects, particularly greenhouse gas emissions. Figure 3.20 shows that the United States area covered by recent droughts (lower panel) is similar to that which prevailed in the 1950s, and also similar to conditions before the reanalysis period such as the "Dust Bowl" era of the 1930s (Box 3.3). Paleoreconstructions of drought conditions for the western United States (upper panel) indicate that recent droughts are considerably less severe and protracted than those that have been estimated for time periods in the twelfth and thirteenth centuries from tree ring data (Cook *et al.*, 2004). Hence, from a frequency/area standpoint, droughts in the recent decades are not particularly outstanding. The causes for these droughts need to be better understood in order to better assess human influences on drought.

southeastern United States; (3) Summer 1988, central/eastern United States; (4) Fall 1995 to Summer 1996, U.S. southern plains; (5) Summer 1998, U.S. southern plains; (6) Summer 1999, eastern United States; (7) 2000 to 2002 western United States/U.S. Great Plains; (8) Spring/summer 2006, centered in Great Plains but widespread.

The droughts discussed above cover various parts of the United States, but droughts are most common in the central and southern Great Plains. Shown in Figure 3.22 is the average summer precipitation for the United States (top) and the seasonal standard deviation for the period 1951 to 2006 (bottom). The largest vari-

While drought can have many definitions, all of the episodes discussed relate to a specific weather pattern that resulted in reduced rainfall, generally to amounts less than 50 percent of average precipitation values. The specific weather pattern in question features an amplified broad-scale high pressure area (ridge) in the troposphere over the affected region (Figure 3.21). Sinking air motion associated with a ridge reduces summertime convective rainfall, results in clear skies with abundant sunshine reaching the surface, and provides for a low-level wind flow that generally prevents substantial moisture advection into the region.

The establishment of a stationary wave pattern in the atmosphere is thus essential for generating severe drought. Such stationary, or blocked atmospheric flow patterns can arise due to mechanisms internal to the atmosphere, and the ensuing droughts can be thought of as due to internal atmospheric processes—so-called unforced variability. However, the longer the anomalous weather conditions persist, the more likely it is to have some stationary forcing acting as a flywheel (*i.e.*, as a source for inertia) to maintain the anomalies.

The droughts discussed above can be distinguished by their duration, with longer lasting events more likely involving forcing of the atmosphere. The atmosphere does not have much heat capacity, and its "memory" of past conditions is relatively short (on the order of a few weeks). Hence, the forcing required to sustain a drought over seasons or years would be expected to lie outside of the atmospheric domain; an obvious possibility with greater heat capacity (and hence a longer "memory") is the ocean. Therefore, most studies have assessed the ability of particular ocean sea surface temperature patterns to generate the atmospheric wave pattern that would result in tropospheric ridges in the observed locations during drought episodes.

Namias (1983) pointed out that the flow pattern responsible for Great Plains droughts, with a ridge over the central United States, also includes other regions of ridging, one in the East Central Pacific and the other in the East Central Atlantic. As described in Chapter 2 and Section 3.1, these teleconnections represent a standing Rossby wave pattern. Using 30 years of data, Namias showed that if the "tropospheric high pressure center in the Central Pacific is strong, there is a good probability of low heights along the West Coast and high heights over the Plains" (Namias, 1983). This further suggests that the cause for the stationary ridge is not completely local, and may have its origins in the Pacific.

Droughts in the western United States are also associated with an amplified tropospheric ridge, which is further west than for Great Plains droughts and in winter displaces storm tracks north of the United States/Canadian border. In winter, the ridge is also associated with an amplified Aleutian Low in the North Pacific, and this has been associated with forcing from the tropical eastern Pacific in conjunction with El Niño events (*e.g.*, Namias, 1978), whose teleconnection and resulting U.S. climate pattern has been discussed in Section 3.1.

Summer Precipitation Climatology

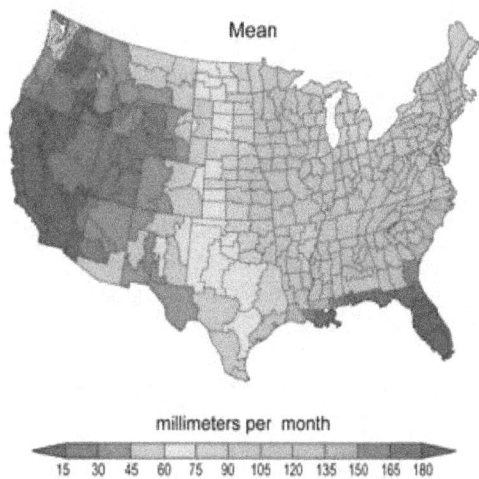

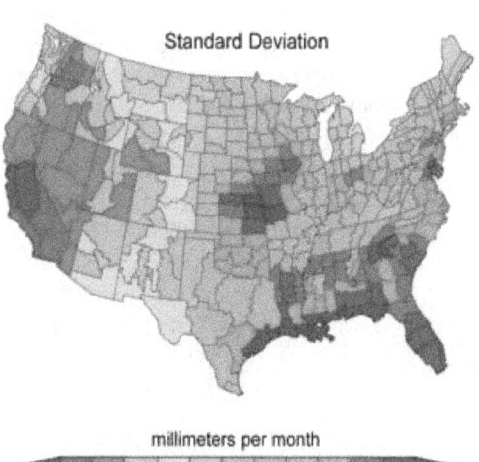

Figure 3.22 Climatological average (top) and standard deviation (bottom) of summer (June-July-August) seasonally-averaged precipitation over the continental United States for the period 1951 to 2006. Contour intervals are (a) 15 millimeters per month and (b) 3 millimeters per day (adopted from Ting and Wang, 1997). Data is the NOAA Climate Division dataset.

Could ENSO also be responsible for warm-season droughts? Trenberth *et al.* (1988) and Trenberth and Branstator (1992) suggested, on the basis of observations and a simplified linear model of atmospheric wave propagation, that colder sea surface temperatures in the tropical eastern Pacific (equatorward of 10°N), the La Niña phase of ENSO, in conjunction with the displacement of warmer water and the Intertropical Convergence Zone (ITCZ) northward

Warm conditions in the Indian Ocean/ West Pacific region are capable of instigating drought in the United States year round but especially in spring.

in that same region (15° to 20°N), led to the amplified ridging over the United States in the spring of 1988. While this was the leading theory at the time, the general opinion now is that most of the short-term summer droughts are more a product of initial atmospheric conditions (Namias, 1991; Lyon and Dole, 1995; Liu *et al.*, 1998; Bates *et al.*, 2001; Hong and Kalnay, 2002) amplified by the soil moisture deficits that arise in response to lack of precipitation (Wolfson *et al.*, 1987; Atlas *et al.*, 1993; Hong and Kalnay, 2002).

For droughts that occur for longer periods of time, various possibilities have been empirically related to dry conditions over specific regions of the United States and Canada. Broadly speaking, they are associated with the eastern tropical Pacific (La Niñas in particular); the Indian Ocean/West Pacific; the North Pacific; and (for the eastern United States) the western Atlantic Ocean. Cool conditions in the eastern tropical Pacific have been related to annual U.S. droughts in various studies (Barlow *et al.*, 2001; Schubert *et al.*, 2004, Seager *et al.*, 2005), although they are more capable of influencing the U.S. climate in late winter when the average atmospheric state is more conducive to allowing an extratropical influence (Newman and Sardeshmukh, 1998; Lau *et al.*, 2006). Warm conditions in the Indian Ocean/West Pacific region are capable of instigating drought in the United States year round (Lau *et al.*, 2006) but especially in spring (Chen and Newman, 1998). Warmer conditions in the North Pacific have

been correlated with drought in the Great Plains (Ting and Wang, 1997) and the U.S. Northeast (Barlow *et al.*, 2001), although modeling studies often fail to show a causal influence (Wolfson *et al.*, 1987; Trenberth and Branstator, 1992; Atlas *et al.*, 1993). The North Pacific SST changes appear to be the result of atmospheric forcing, rather than the reverse; therefore, if they are contributing to drought conditions, they may not be the cause of the initial circulation anomalies. Alexander *et al.* (2002) concluded from Global Circulation Model (GCM) experiments that roughly one-quarter to one-half of the change in the dominant pattern of low frequency variability in the North Pacific sea surface temperatures during winter was itself the result of ENSO, which helps intensify the Aleutian Low and increases surface heat fluxes (promoting cooling).

Sea surface temperature perturbations downstream of North America, in the North Atlantic, have occasionally been suggested as influencing some aspects of U.S. drought. For example, Namias (1983) noted that the wintertime drought in the western United States in 1977, one of the most extensive far western droughts in recent history, appeared to be responsive to a downstream deep trough over the eastern United States. Warmer sea surface temperatures in the western North Atlantic have the potential to intensify storms in that region. Conversely, colder sea surface temperatures in summer can help intensify the ridge (*i.e.*, the "Bermuda High") that exists in that region. Namias (1966) suggested that such a cold water regime played an integral part in the U.S. Northeast spring and summer drought of 1962 to 1965, and Schubert *et al.* (2004) find Atlantic SST effects on the Dust Bowl, while multi-decadal swings between wet and dry periods over the United States as a whole have been statistically linked with Atlantic SST variations of similar time-scale (McCabe *et al.*, 2004; Figure 3.5).

In Mexico, severe droughts during the reanalysis period were noted primarily in the 1950s, and again in the 1990s. The 1990s time period featured seven consecutive years of drought (1994 to 2000). Similar to the United States, droughts in Mexico have been linked to tropospheric ridges that can affect northern Mexico, and also to ENSO. However, there are additional

factors tied to Mexico's complex terrain and its strong seasonal monsoon rains. Mexican rainfall in the warm season is associated with the North American Monsoon System (NAMS), which is driven by solar heating from mid-May into July. Deficient warm season rainfall over much of the country is typically associated with El Niño events. La Niña conditions often produce increased rainfall in southern and northeastern Mexico, but have been associated with drought in northwestern Mexico (Higgins *et al.*, 1999). During winter and early spring, there is a clear association with the ENSO cycle (*e.g.*, Stahle *et al.*, 1998), with enhanced precipitation during El Niño events associated with a strengthened subtropical jet that steers storms to lower latitudes and reduced rainfall with La Niñas when the jet moves poleward.

Therefore, the occurrence of drought in Mexico is heavily dependent on the state of the ENSO cycle, or its teleconnection to the extratropics, and on solar heating variations. In the warm season there is often an out-of-phase relationship between southern and northern Mexico, and between spring and summer, dependent on the phasing of the NAMS (Therrell *et al.*, 2002). These aspects make attribution of recent droughts difficult. For example, the consecutive drought years from 1994 to 2000 occurred over several different phases of ENSO, suggesting multiple causes including El Niño conditions for warm season drought through 1998, the possible influence of Indian Ocean/West Pacific warming during the subsequent La Niña phase, and internal atmospheric variability.

Because a large proportion of the variance of drought conditions over North America is unrelated to sea surface temperature perturbations, it is conceivable that when a severe drought occurs it is because numerous mechanisms are acting in tandem. This was the conclusion reached in association with the recent U.S. drought (1999 to 2005) that affected large areas of the southern, western and central United States. During this time, warm conditions prevailed over the Indian Ocean/West Pacific region along with La Niña conditions in the eastern tropical Pacific—influences from both regions working together may have helped intensify and/or prolong the annual droughts (Hoerling and Kumar, 2003; Lau *et al.*, 2006).

3.5.3.2 HUMAN INFLUENCES ON NORTH AMERICAN DROUGHT SINCE 1951

To the extent that ENSO cycle variations (La Niñas in particular) are the cause of drought in the United States, it is difficult to show that they are related to greenhouse gas forcing. While some studies (*e.g.*, Clement *et al.*, 1996) have suggested that La Niña conditions will be favored as climate warms, in fact more intense El Niño events have occurred since the late 1970s, perhaps due at least in part to anthropogenic warming of the eastern equatorial Pacific (Mendelssohn *et al.*, 2005). There is a tendency in model projections for the future greenhouse-gas warmed climate to indicate an average shift towards more El Niño-like conditions in the tropical eastern Pacific Ocean, including the overlying atmospheric circulation; this latter aspect may already be occurring (Vecchi and Soden, 2007). With respect to the human influence on ENSO variability, Merryfield (2006) surveyed 15 coupled atmosphere-ocean models and found that for future projections, almost half exhibited no change, five showed reduced variability, and three increased variability. Hence, to the extent that La Niña conditions are associated with drought in the United States, there is no indication that they have been or will obviously be influenced by anthropogenic forcing.

However, given that SST changes in the Indian Ocean/West Pacific are a factor for long-term U.S. drought, a somewhat different story emerges. Shown in Figure 3.23 are the SST anomalies in this region, as well as the tropical central-eastern Pacific (Lau *et al.*, 2006). As noted with respect to the recent droughts, the Indian Ocean/West Pacific region has been consistently warm when compared with the 1971 to 2000 sea surface temperature climatology. What has caused this recent warming?

The effect of more frequent El Niños alone results in increased temperatures in the Indian Ocean, acting through an atmospheric bridge that alters the wind and perhaps the cloud field in the Indian Ocean region (Klein *et al.*, 1999; Yu and Rienecker, 1999; Alexander *et al.*, 2002; Lau and Nath, 2003); an oceanic bridge between the Pacific and the Indian Ocean has also been modeled (Bracco *et al.*, 2007). This effect could then influence droughts over the

Because a large proportion of the variance of drought conditions over North America is unrelated to sea surface temperature perturbations, it is conceivable that when a severe drought occurs it is because numerous mechanisms are acting in tandem.

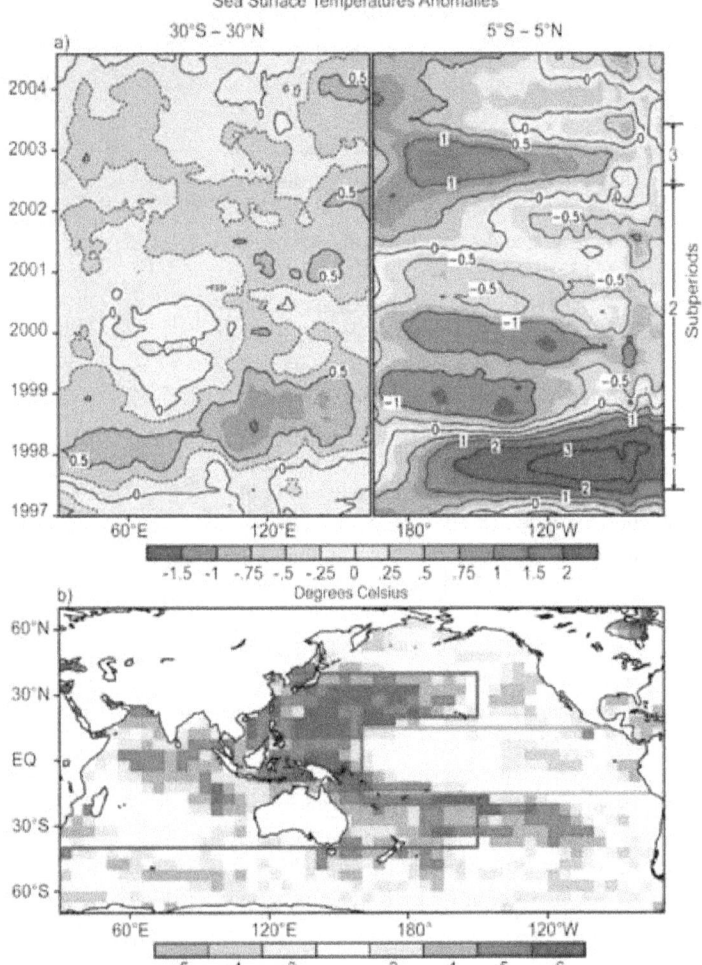

Figure 3.23 Top panel: Sea surface temperature anomalies relative to the period 1970 through 2000 as a function of year in the Indian Ocean/West Pacific (left) and central-eastern Pacific (right) (from Lau *et al.*, 2006). Bottom panel: Number of 12-month periods in June 1997 to May 2003 with SST anomalies at individual 5° latitude by 5° longitude rectangles being above normal (red shading) or below normal (blue shading) by more than one-half of a standard deviation (*i.e.* one-half the strength of the expected variability).

pogenic forcing for this region and is outside the range expected from natural variability, as judged by coupled atmosphere-ocean model output of the CMIP simulations (Hegerl *et al.*, 2007). The comparison of the observed warm pool SST time series with those of the CMIP simulations in Section 3.2.2 indicates that it is very likely that the recent warming of SSTs over the Indian Ocean/West Pacific region is of human origins.

The possible poleward expansion of the subtropical region of descent of the Hadley Circulation is an outcome that is favored by models in response to a warming climate (IPCC, 2007a). It would transfer the dry conditions of northern Mexico to the U.S. Southwest and southern Great Plains; Seager *et al.* (2007) suggest that may already be happening, and is associated with drought in the southwestern United States. Additional observations and modeling improvements will be required to assess the likelihood of its occurrence with greater confidence.

An additional impact of greenhouse warming is a likely increase in evapotranspiration during drought episodes because of warmer land surface temperatures. It was noted in the discussion of potential causes that reduced soil moisture from precipitation deficits helped sustain and amplify drought conditions, as the surface radiation imbalance increased with less cloud cover, and sensible heat fluxes increased in lieu of latent heat fluxes. This effect would not have initiated drought conditions but would be an additional factor, one that is likely to grow as climate warms. For example, drier conditions have been noted in the northeast United States despite increased annual precipitation, due to a century-long warming (Groisman *et al.*, 2004); this appears to be true for Alaska and southern and western Canada as well (Dai *et al.*, 2004). Droughts in the western United States also appear to have been influenced by increasing temperature (Andreadis and Lettenmaier, 2006; Easterling *et al.*, 2007). The areal extent of forest fires in Canada has been high since

United States in the summer after an El Niño, as opposed to the direct influence of La Niña (Lau *et al.*, 2005).

Nevertheless, as shown in Figure 3.23, the warming in the Indian Ocean/West Pacific region has occurred over different phases of the ENSO cycle, making it less likely that the overall effect is associated with it. Hoerling and Kumar (2003) note that "the warmth of the tropical Indian Ocean and the western Pacific Ocean was unsurpassed during the twentieth century"; the region has warmed about 1°C (1.8°F) since 1950. That is within the range of warming projected by models due to anthro-

1980 compared with the previous 30 years and Alaska experienced record forest fire years in 2004 and 2005 (Soja *et al.*, 2007). Hence, by adding additional water stress global warming can exacerbate naturally occurring droughts, in addition to influencing the meteorological conditions responsible for drought.

A further suggestion of the increasing role played by warm surface temperatures on drought is given in Figure 3.24. A diagnosis of conditions during the recent U.S. Southwest drought is shown, with contours depicting the atmospheric circulation pattern based on reanalysis data, and shading illustrating the surface temperature anomaly (top) and precipitation anomaly (bottom). High pressure conditions prevailed across the entire continent during the period, acting to redirect storms far away from the region. Continental-scale warmth during 1999 to 2004 was also consistent with the anthropogenic signal. It is plausible that the regional maximum in warmth seen over the Southwest during this period was in part a feedback from the persistently below normal precipitation, together with the anthropogenic signal. Overall, the warmth associated with this recent drought has been greater than the warmth observed during the 1950s drought in the Southwest (Figure 3.21), likely augmenting its negative impacts on water resource and ecologic systems compared to the earlier drought.

Breshears *et al.* (2005) estimated the vegetation die-off extent across southwestern North America during the recent drought. The combination of drought with pine bark beetle infestation resulted in more than a 90 percent loss in Piñon pine trees in some areas. They noted that such a response was much more severe than during the 1950s drought, arguing that the recent drought's greater warmth was the material factor explaining this difference.

Current understanding is far from complete concerning the origin of individual droughts. While the assessment discussed here has emphasized the apparently random nature of short-term droughts, a product of initial conditions which then sometimes develop rapidly into strong tropospheric ridges, the relationships of such phenomena to sea surface temperature patterns, including the ENSO cycle, are still

being debated. The ability of North Atlantic sea surface temperatures to influence the upstream circulation still needs further examination in certain circumstances, especially with respect to droughts in the eastern United States. The exact mechanisms for influencing Rossby wave

1999-2004 Annual Composite

Temperature

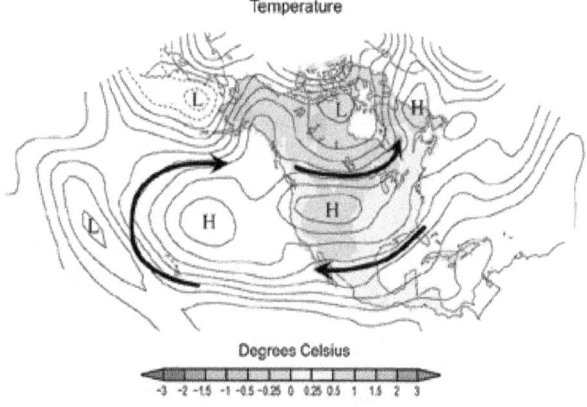

Degrees Celsius

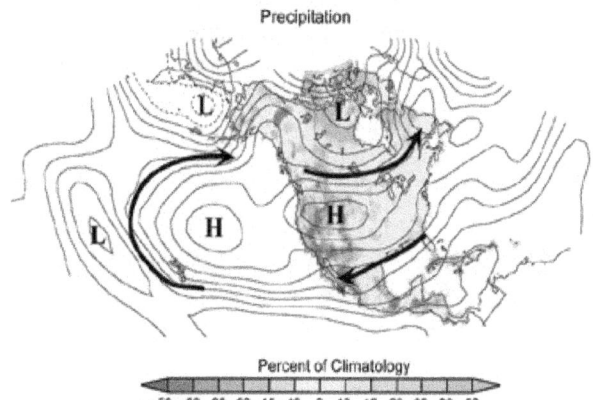

Precipitation

Percent of Climatology

Figure 3.24 Observed climate conditions averaged for 1999 to 2004 during a period of severe southwestern U.S. drought. The 500 millibar height field (contours, units meters) is from the NCEP/NCAR RI reanalysis. The shading indicates the five-year average anomaly of the surface temperature (top) and precipitation (bottom). The surface temperature and precipitation are from independent observational datasets. Anomalous High and Low Pressure regions are highlighted. Arrows indicate the anomalous wind direction, which circulates around the High (Low) Pressure centers in a clockwise (counterclockwise) direction.

development downstream, including the role of transient waves relative to stationary wave patterns, will undoubtedly be the subject of continued research. The Hadley Cell response to climate change, as noted above, is still uncertain. Also, while some modeling studies have emphasized the role played by surface

The severity of both short- and long-term droughts has likely been amplified by local greenhouse gas warming in recent decades.

soil moisture deficits in exacerbating these droughts, the magnitude of the effect is somewhat model-dependent, and future generations of land-vegetation models may act somewhat differently.

Given these uncertainties, it is concluded from the above analysis that, of the severe droughts that have impacted North America over the past five decades, the short-term (monthly-to-seasonal) events are most likely to be primarily the result of initial atmospheric conditions, subsequently amplified by local soil moisture conditions, and in some cases initiated by teleconnection patterns driven in part by SST anomalies. For the longer-term events, the effect of steady forcing through sea surface temperature anomalies becomes more important. Also, the accumulating greenhouse gases and global warming have increasingly been felt as a causative factor, primarily through their influence on Indian Ocean/West Pacific temperatures, conditions to which North American climate is sensitive. The severity of both short- and long-term droughts has likely been amplified by local greenhouse gas warming in recent decades.

CHAPTER 4

Recommendations

Convening Lead Author: Randall Dole, NOAA/ESRL

Lead Authors: Martin Hoerling, NOAA/ESRL; Siegfried Schubert, NASA/GMAO

Contributing Authors: Phil Arkin, Univ. of Maryland; James Carton, Univ. of Maryland; Gabi Hegerl, Univ. of Edinburgh; David Karoly, Univ. of Melbourne; Eugenia Kalnay, Univ. of Maryland; Randal Koster, NASA/GMAO; Arun Kumar, NOAA/CPC; Roger Pulwarty, NOAA/CPO; David Rind, NASA/GISS

RECOMMENDATIONS

This Chapter discusses steps needed to improve national capabilities in climate analysis, reanalysis, and attribution in order to better address key issues in climate science and to increase the value of this information for applications and decision making. Limitations, gaps in current capabilities, and opportunities for improvement that have been identified in earlier chapters, together with several related studies and reports, provide the primary foundations for the findings and recommendations provided here. The overarching goal is to provide high-level recommendations to improve national capabilities in climate analyses, reanalyses, and attribution in order to increase their value for scientific and practical applications.

4.1 NEED FOR A SYSTEMATIC APPROACH TO CLIMATE ANALYSIS AND REANALYSIS

Climate analyses and reanalyses are being used in an increasing range of practical applications in sectors such as energy, agriculture, water resources, and insurance.

As discussed throughout this report, reanalysis products have played a major role in advancing climate science and have supported numerous applications, including: monitoring climate and comparing current conditions with those of the past; providing initial conditions required for climate model predictions; supporting research on climate variability and change; enabling more reliable climate attribution; and providing benchmarks for evaluating climate models. Climate analyses and reanalyses are also being used in an increasing range of practical applications in sectors such as energy, agriculture, water resources, and insurance (*e.g.,* Schwartz and George, 1998; Pryor *et al.,* 2006; Challinor *et al.,* 2005; Pulwarty, 2003; Pinto *et al.,* 2007).

These are important benefits. However, there are limitations in climate analysis and reanalysis products that presently constrain their value. Perhaps the largest constraint for climate applications is that, while the model and data assimilation system remain the same over the reanalysis period, the observing system does not, and this can lead to false trends, jumps, and other uncertainties in climate records (Arkin *et al.,* 2004; Simmons *et al.,* 2006; Bengtsson *et al.,* 2007).

Another constraint is the limited length of present reanalysis records, which now extend back to 1948, at most. Extending reanalysis back over a century or longer would improve descriptions and attribution of causes of important climate variations, such as the pronounced warm interval in the 1930s and 1940s, the Dust Bowl drought over much of the United States in the 1930s, and multi-decadal climate variations. International efforts such as the Global Climate Observing System (GCOS, 2004) and Global Earth Observation Systems of Systems (GEOSS, 2005) have identified the need for reanalysis datasets extending as far back as possible in order to compare recent and projected climate changes with those of the past.

The development of current climate analysis and reanalysis activities, while encouraging and beneficial, is occurring without clear coordina-

tion at national interagency levels, which may result in less than optimal progress and the inability to ensure a focus on problems of greatest scientific and public interest. Currently, no U.S. agency is charged with primary responsibility to ensure that the Nation has an ongoing capability in climate analysis or reanalysis, putting the sustainability of national capabilities at some risk.

The following recommendations focus on the value, needs, and opportunities for climate analysis and reanalysis in providing consistent descriptions and attribution of past climate variability and change, and in supporting applications and decision making at relevant national, regional and local levels. The recommendations point to the necessity for improved coordination between U.S. agencies and with international partners to develop an ongoing climate analysis and systematic reanalysis capacity that would address an increasing range of scientific and practical needs.

4.2 RECOMMENDATIONS FOR IMPROVING FUTURE CLIMATE ANALYSES AND REANALYSES

- **To better detect changes in the climate system, improve the quality and consistency of the observational data and reduce effects of observing system changes.**

As discussed in this Product, changes in observing systems (for example, the advent of comprehensive satellite coverage in the late 1970s), create significant uncertainties in the detection of true climate variations and trends over multiple decades. To reduce these uncertainties, closer collaborations are needed between observational and reanalysis communities to improve the existing global database of Earth system observations (Schubert *et al.,* 2006). Priorities include improving quality control, identification and correction of observational bias and other errors, the merging of various datasets, data recovery, improving the handling of metadata (that is, information describing how, when, and by whom a particular set of data was collected, content and structure of records, and their management through time), and developing and testing techniques to more effectively adjust to changes in observing systems (Dee, 2005).

This recommendation resonates with recommendations from other reports, including CCSP Synthesis and Assessment Product 1.1, which focuses on steps for understanding and reconciling differences in temperature trends in the lower atmosphere (CCSP, 2006). That report stated:

> Consistent with Key Action 24 of GCOS (2004) and a 10 Year Climate Target of GEOSS (2005), efforts should be made to create several homogeneous atmospheric reanalyses. Particular care needs to be taken to identify and homogenize critical input climate data, and to more effectively manage large-scale changes in the global observing system to avoid non-climatic influences (CCSP, 2006).

Recent World Meteorological Organization (WMO) reports emphasize the need for ongoing climate analyses and periodic reanalyses as critical parts of the Global Climate Observing System (GCOS) , *e.g.*, GCOS (2003, 2004), Simmons *et al.* (2006), Trenberth *et al.* (2006). GCOS (2004) states that "Parties are urged to give high priority to establishing a sustained capacity for global climate reanalysis, and to develop improved methods for such reanalysis, and to ensure coordination and collaboration among Centers in conducting reanalyses".

Data quality control and expanding the use of available observations will be crucial to this effort. Significant gains are possible for both satellite and conventional observations (Arkin *et al.*, 2004). More research is required to understand biases in individual satellite data collections, to account for different resolutions and sensor measurements, and to minimize the impact of transitions between satellite missions, which may lead to data gaps or to apparent discontinuities if the satellite measurements are not cross-calibrated, *e.g.*, by comparing measurements obtained over an overlapping time period for the missions. In addition, early satellite data from the late 1960s and 1970s need further quality control and processing before they can be used effectively in reanalyses. Dedicated efforts are required to determine the full effects

of changes in the observing systems, to focus on bias-corrected observations, and to assess remaining uncertainties in trends and estimates of variability. Observing System Experiments (OSEs) that consider the effects of inclusion or removal of particular data can be helpful in identifying and reducing possible harmful impacts of changes in observing systems.

- **Develop analysis methods that are optimized for climate research and applications. These methods should include uncertainty estimates for all reanalysis products.**

As discussed in Chapter 2, data assimilation techniques used in initial climate reanalyses were developed from methods optimized for use in numerical weather prediction. The primary goal of numerical weather prediction is to produce the best forecast. True four-dimensional data assimilation methods (using data that includes observations from before and after the analysis time, which is the start time of the forecast) have been developed for numerical weather prediction. However, the requirements for weather forecasts to be ready within a short time frame (typically within a few hours after the analysis time) result in observational data obtained after the beginning of the forecast cycle either not being assimilated at all or treated differently from observations obtained before or at the analysis time. The strong constraints placed by the needs for timely forecasts also substantially limit the capability of analyses to use the full historical observational database, which may not be collected until long after the forecast is completed.

These constraints are not relevant for

There are many
applications of
reanalyses, and it is
likely that different
scientific approaches
will be required to
optimally address
particular problems.

climate analyses, and modification of current data assimilation methods is needed to improve representations of long-term trends and variability (Arkin *et al.*, 2004). Further, many potentially available observations, including numerous satellite, surface temperature, and precipitation observations, could not be effectively assimilated within the first atmospheric reanalyses due to limitations of the models and assimilation techniques, and because some data were not available when the reanalyses were conducted (Kalnay *et al.*, 1996). Advances in data assimilation that have occurred since the pioneering reanalysis projects enable better and more complete use of these additional observations.

To produce reanalyses that better serve climate research and applications, it will be essential to develop methods to more effectively use the wealth of information provided by diverse Earth observations, reduce the sensitivity of the data assimilation to changes in the observing system, and provide estimates of remaining uncertainties in reanalysis products. A major emphasis for efforts should be on the modern satellite era, essentially 1979 to present, during which time the number and diversity of observational data have expanded greatly but have yet to be fully utilized. An important development that should help to achieve this goal is the national Earth System Modeling Framework (ESMF, <http://www.esmf.ucar.edu>). The ESMF is a collaborative effort between NASA, NOAA, the National Science Foundation, and the Department of Energy that is developing the overall organization, infrastructure, and low-level utilities required to allow the interchange of models, model sub-components, and analysis systems. ESMF should greatly expand the ability of scientists outside the main data assimilation centers (*e.g.*, in universities and other organizations) to accelerate progress in addressing key challenges toward improving the analyses.

There are many applications of reanalyses, and it is likely that different scientific approaches will be required to optimally address particular problems. For example, if the primary goal is to optimize the detection of climate trends, particular care must be given to minimizing effects of changing observing systems so as to ensure the highest quality analysis over an extended time

period. In this case, an appropriate reanalysis strategy may be to use a subset of very high quality data that is available continuously, or nearly continuously, over as long a period as feasible. Conversely, if the primary goal is to perform detailed studies of processes at high space and time resolutions, the most accurate analysis at any given time may be preferred. Here, the best strategy may be to take advantage of all available observations. In any case, uncertainties in the analyses and their implications should be appropriately documented.

Alternative data assimilation methods should be explored for their potential benefits. One alternative that is being examined intensively, ensemble data assimilation, shows considerable promise in addressing a wide range of problems. This technique uses multiple model predictions (called an "ensemble") to estimate where errors may be particularly large or small at a given time. This time- and location-dependent uncertainty information is then incorporated into the analysis (Houtekamer and Mitchell, 1998; Whitaker and Hamill, 2002). This approach provides estimates of uncertainties in the full range of reanalysis products (including, for example, the components of the water cycle). Ensemble data assimilation is becoming more economical with the development of innovative methods to take advantage of massively parallel computing (Ott *et al.*, 2004). In addition, ensemble-based approaches are being developed that explicitly account for model error (Zupanski and Zupanski, 2006), thereby providing a potentially important step toward better estimating analysis uncertainties.

- **To improve the description and understanding of major climate variations that occurred prior to the mid-twentieth century, develop the longest possible consistent record of past climate conditions.**

For many applications, the relatively short period encompassed by initial reanalyses is a very important constraint. Current reanalysis datasets extend back to the mid-twentieth century at most. As a consequence, many climate variations of great societal interest, such as the prolonged Dust Bowl drought of the 1930s, are not included in present reanalyses, increasing

uncertainties in both their descriptions and causes.

Recent research has demonstrated that a reanalysis through the entire twentieth century, and perhaps earlier, is feasible using only surface pressure observations (Whitaker *et al.*, 2004; Compo *et al.*, 2006). Extending reanalysis back over a century or longer would be of great value in improving descriptions and attribution of causes of important climate variations such as the pronounced warm interval in the 1930s and 1940s, the Dust Bowl drought, and other multi-decadal climate variations. International efforts such as the GCOS and GEOSS have identified the need for reanalysis datasets extending as far back as possible to compare the patterns and magnitudes of recent and projected climate changes with past changes (GCOS, 2004; GEOSS, 2005). Such reanalysis datasets should also enable researchers to better address issues on the range of natural variability of weather and climate and increase understanding of how El Niño-Southern Oscillation and other climate patterns alter the behavior of extreme events.

Alternative assimilation methods should also be evaluated for obtaining maximum information for estimating climate variability and trends from very sparse observations and from surface observations alone, where observational records are available over much longer periods than other data sources. Ensemble data assimilation methods have already shown considerable promise in this area (Ott *et al.*, 2004; Whitaker *et al.*, 2004; Compo *et al.*, 2006; Simmons *et al.*, 2006), and, as mentioned previously, also provide estimates of analysis uncertainty. Improved methods of estimating and correcting observational and model errors, recovery of historical observations, and the development of optimal, consistent observational datasets will also be required in this effort.

- **To improve decision support, produce future climate reanalysis products at finer space scales (*e.g.*, resolutions of 10 miles rather than 100 miles) and emphasize products that are most relevant for applications, such as surface temperatures, winds, cloudiness, and precipitation.**

For many applications, the value of the initial reanalysis products has been constrained by their relatively coarse horizontal resolution (200 kilometers or approximately 120 miles). For many users, improved representation of the water cycle (inputs, storage, outputs) is a key need. In addition, land-surface processes are important for both surface energy (temperature) and water balance, with land cover and land use becoming increasingly important at smaller scales. These processes should be research focus areas for future improvements.

Within the United States, one step forward in addressing these issues is the implementation of NASA's new reanalysis project (MERRA, see Box 2.2 for a detailed description), which will provide global reanalyses at approximately 50-kilometer (about 30-mile) resolution and has a focus on providing improved estimates of the water cycle <http://gmao.gsfc.nasa.gov/research/merra/>. Another important step forward is the completion of the North American Regional Reanalysis, or NARR (Mesinger *et al.*, 2006). While this is a regional analysis for North America and adjacent areas, rather than global reanalysis, it is at considerably higher resolution than the global reanalyses with a grid spacing of 32 kilometers (about 20 miles). Importantly, NARR also incorporates significant advances in modeling and data assimilation that were made following the initial global reanalysis by NOAA and the National Center for Atmospheric Research (Kalnay *et al.*, 1996), including the ability to assimilate precipitation observations. This has resulted in substantial improvements in precipitation analyses over the contiguous United States as well as improvements in near-surface temperatures and wind fields (Mesinger *et al.*, 2006). While advances are impressive, early studies show that further improvements are

Extending reanalysis back over a century or longer would be of great value in improving descriptions and attribution of causes of important climate variations such as the pronounced warm interval in the 1930s and 1940s, the Dust Bowl drought, and other multi-decadal climate variations.

needed to accurately represent the complete water cycle (*e.g.*, Nigam and Ruiz-Barradas, 2006). The ability to improve analyses of key surface variables and the water cycle therefore remain as important challenges.

- **Develop new national capabilities in analysis and reanalysis that focus on variables that are of high relevance to policy and decision support. Such variables include those required to monitor changes in the carbon cycle and to understand interactions among Earth system components (atmosphere, ocean, land, cryosphere, and biosphere) that may lead to accelerated or diminished rates of climate change.**

Initial reanalyses focused on reconstructing past atmospheric conditions. For both scientific and practical purposes, there is a strong need to consider other Earth system components, such as the ocean, land, cryosphere, and biosphere, as well as variables that are of interest for climate but are of less immediate relevance for weather prediction (*e.g.*, related to the carbon cycle). As discussed in Chapter 2, such efforts are ongoing for ocean and land data assimilation but are still in relatively early stages. The long-term goal should be to move toward ongoing analyses and periodic reanalyses of major Earth system components relevant to climate variability and change.

Future climate analyses and reanalyses should incorporate additional climate system components that are relevant for decision making and policy development, for example, a carbon cycle to aid in identifying changes in carbon emissions sources and removal processes. A reanalysis of the chemical state of the atmosphere would improve monitoring and understanding of air quality in a changing climate, aerosol-climate interactions, and other key policy-relevant issues. Initial attempts at coupling of climate system components, *e.g.*, ocean-atmosphere reanaly-

sis, should be fostered, with a long-term goal being to develop an integrated Earth system analysis (IESA) capability that includes interactions among the Earth system components (atmosphere, ocean, land, snow and ice, and biological systems).

An IESA would provide the scientific community, resource managers, decision makers, and policy makers with a high quality, internally consistent, continuous record of the Earth system that can be used to identify, monitor, and assess changes in the system over time. Developing an IESA would also contribute to improved descriptions and understanding of the coupled processes that may produce rapid or accelerated climate changes, for example, from high-latitude feedbacks related to changes in sea ice or melting of permafrost that may amplify an initial warming due to natural or anthropogenic causes. Key processes include: atmosphere-ocean interactions for physical and biogeochemical processes; climate feedbacks from snow and ice processes; carbon cycle feedbacks; and atmosphere-land-biosphere interactions.

To achieve an IESA will require a sustained capacity to assimilate current and planned future observations from diverse platforms into Earth system models. This approach will be essential for realizing the full value of investments in current and proposed future observing systems within GEOSS, as it provides the means of integrating diverse datasets together to obtain a unified, physically consistent description of the Earth system. It would also take advantage of rapid advances in Earth system modeling, while providing the ability to evaluate models used for attribution and climate predictions and projections.

Recent efforts have shown the feasibility of extending initial atmospheric analyses beyond traditional weather variables. For example, the European Union has funded a project, the Global Environment Monitoring System (GEMS), that is incorporating satellite and *in situ* data (data collected at its original location) to develop an analysis and forecast capability for atmospheric aerosols, greenhouse gases, and reactive gases (Hollingsworth *et al.*, 2005). The GEMS operational system will be an extension

Future climate analyses and reanalyses should incorporate additional climate system components that are relevant for decision making and policy development, for example, a carbon cycle to aid in identifying changes in carbon emissions sources and removal processes.

of current weather data assimilation capabilities, with implementation planned for 2009. The main users of the GEMS Project are intended to be policy-makers, operational regional air quality and environmental forecasters, and the scientific community. GEMS will support operational regional air-quality and "chemical weather" forecast systems across Europe. Part of the motivation for this project is to provide improved alerts for events such as the 2003 heat waves in Western Europe that led to at least 22,000 deaths, mostly due to heat stress, but also connected to poor air quality (Kosatsky, 2005). GEMS will generate a reanalysis of atmospheric dynamics and composition, and state-of-the-art estimates of the emissions sources and removal processes as well as how gases and aerosols are transported across continents. These estimates are designed to meet key information requirements of policy-makers, and to be relevant to the Kyoto and Montreal Protocols and the United Nation Convention on long-range trans-boundary air pollution (Hollingsworth *et al.*, 2005).

Within the United States, NOAA has developed plans to use a fully coupled atmosphere-land-ocean-sea ice model for its next generation of global reanalysis, extending over the period 1979 to 2008 (S. Saha, personal communication, 2007). This reanalysis is based on the NOAA-NCEP Climate Forecast System (CFS) model (Saha *et al.*, 2006). While the component analyses will be performed separately through independent atmosphere, land, ocean, and sea ice data assimilation systems, the use of a coupled model provides consistent initial estimates for all variables that is an important step toward a fully coupled Earth system analysis. Current plans are to begin production and evaluation of this reanalysis in 2008. This global atmosphere-ocean reanalysis would provide important advances on a number of fronts, taking advantage of improvements in modeling, data assimilation, and computing that have occurred since the pioneering NCEP-NCAR reanalysis. Atmospheric resolution will also be greatly increased, from approximately 200 kilometers (120 miles) in the earlier version to 30 to 40 kilometers (around 20 to 25 miles) in the new version. In addition to atmospheric, ocean, and land data assimilation, significant new efforts are examining the use of data as-

simulation techniques to analyze other aspects of the Earth system, with one important focus being to better represent and identify gas and aerosol emissions sources and removal processes in the atmospheric carbon cycle (Peters *et al.*, 2005).

- **Develop a more coordinated, effective, and sustained national capability in analysis and reanalysis to support climate research and applications.**

Without a clear and systematic institutional commitment, future efforts in climate analysis and reanalysis are likely to be *ad hoc*, and are unlikely to result in high quality, sustained, cost-effective products. Developing a national capability in climate (and Earth system) analysis and reanalysis will be essential to achieving key CCSP objectives and, in particular, CCSP Goal 1: "Improve knowledge of the Earth's past and present climate and environment, including its natural variability, and improve understanding of the causes of climate variability and change" (CCSP, 2003).

This idea was first highlighted over 15 years ago in a National Research Council report (NRC, 1991) that outlined a strategy for a focused national program on data assimilation for the Earth system. A key recommendation of that report was that "A coordinated national program should be implemented and funded to develop consistent, long term assimilated datasets ... for the study of climate and global change". This recommendation has been reiter-

Without a clear and systematic institutional commitment, future efforts in climate analysis and reanalysis are likely to be *ad hoc*, and are unlikely to result in high quality, sustained, cost-effective products.

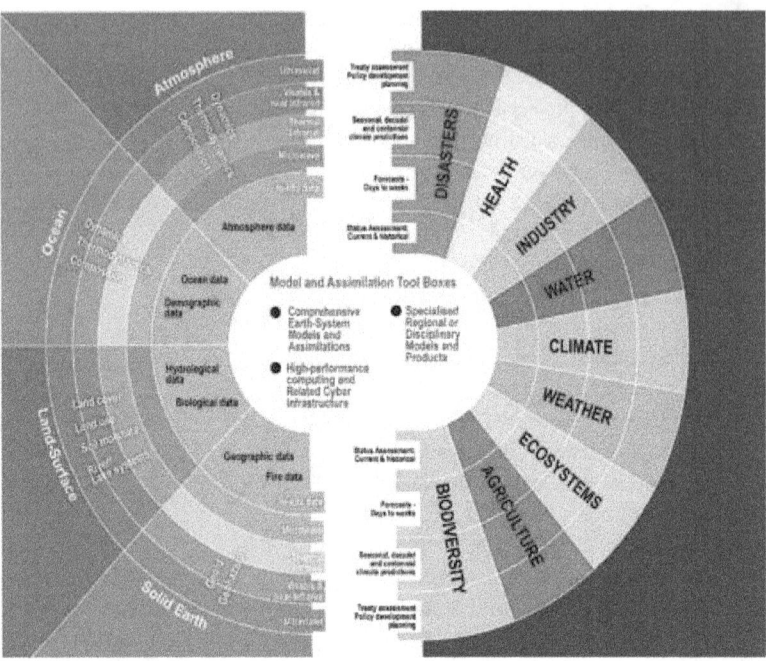

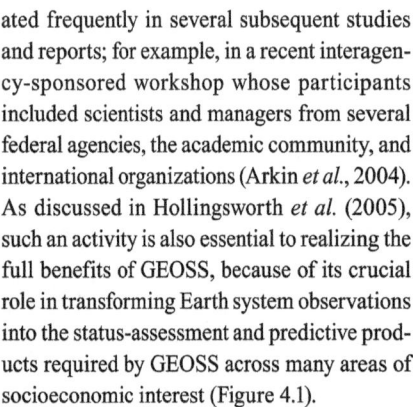

Figure 4.1 From Hollingsworth (2005), based on the GEOSS Implementation Plan (GEOSS, 2005), illustrating the transformation of observations into predictive and current-status information. On the right-hand side are products from an Earth system forecasting system and associated specialized models organized in GEOSS categories of socioeconomic benefits, stratified by the lead time required for products (current status assessments, forecast time-range, long-term studies of reanalysis). On the left-hand side are observational requirements for a comprehensive Earth system model, including *in situ* data as well as current and projected satellite data. In the center are "tool boxes" needed to achieve the transformation from observations into information.

ated frequently in several subsequent studies and reports; for example, in a recent interagency-sponsored workshop whose participants included scientists and managers from several federal agencies, the academic community, and international organizations (Arkin *et al.*, 2004). As discussed in Hollingsworth *et al.* (2005), such an activity is also essential to realizing the full benefits of GEOSS, because of its crucial role in transforming Earth system observations into the status-assessment and predictive products required by GEOSS across many areas of socioeconomic interest (Figure 4.1).

To be truly successful such a program must include multiple agencies, since it requires resources and expertise in a broad range of scientific disciplines and technologies beyond that of any single agency (*e.g.*, atmosphere, ocean, land surface and biology, observations and modeling, measurements, computing, data visualization and delivery). It also will need strong ties with the Earth science user community, to ensure that the analysis and reanalysis products satisfy the requirements of a broad spectrum of users and provide increasing value over time.

4.3 NEED FOR IMPROVED CLIMATE ATTRIBUTION

Recent events underscore the socioeconomic significance of credible and timely climate attribution. For instance, the recent extremely warm year of 2006 in the United States raises questions over whether the probability of occurrence of such warm years has changed, what the factors are contributing to the changes, and how such factors might alter future probabilities of similar or warmer years. Policy and decision makers want to know the answers to these questions because this information can be used within planning and response strategies. What climate processes are responsible for the persistent western U.S. drought and what implications does this have for the future? Planners are assessing the sustainability and capacity of the region for further growth, and the resilience of water resources to climate variations and change is an important factor that they must consider. What processes contributed to the extremely active 2004 and 2005 North Atlantic hurricane seasons as well as to the general increase in hurricane activity in this region since the mid-1990s? Emergency managers want to know the answers to such questions and related implications for the coming years, in order to better prepare for the future.

This Product has identified several outstanding challenges in attribution research that are motivated by observed North American climate variations that occurred during the reanalysis period but have yet to be fully explained. For instance, what is the cause of the so-called summertime "warming hole" over the central United States? The results of Chapter 3 indicate that this pattern is inconsistent with an expected anthropogenic warming obtained from coupled model simulations, although model simulations with specified sea surface temperature variations over the period are able to represent aspects of this pattern. Other forcings resulting from human activities, including by atmospheric aerosols and land use and land cover changes, may play significant roles but their effects have yet to be quantified. From a decision-making perspective, it is important to know whether the absence of summertime warming in the primary grain producing region of the United States is a natural climate variation that may be temporarily offsetting long-term human-induced warming, or whether current climate models contain specific errors that are leading to systematic overestimates of projected warming for this region.

As emphasized in Hegerl *et al.* (2006), to better serve societal interests there is a need to go beyond detecting and attributing the causes of global average surface temperature trends to consider the causes of other climate variations and changes. As detection and attribution studies move toward smaller scales of space and time and consider a broader range of variables, important challenges must be addressed.

4.4 RECOMMENDATIONS FOR IMPROVING CLIMATE ATTRIBUTION CAPABILITIES

- **Develop a national capability in climate attribution to provide regular and reliable explanations of evolving climate conditions relevant to decision making.**

Similar to the present status of U.S. efforts in climate analysis and reanalysis, attribution research is presently supported through a range of agency research programs without clear national coordination (Trenberth *et al.*, 2006). This may limit abilities to address attribution problems of highest scientific or public interest. There are also no clearly designated responsibilities to communicate state-of-science findings on attribution. Therefore, the public and media are often exposed to an array of opinions on causes for observed climate events, with diametrically opposed views sometimes expressed by different scientists from within the same agency. In many cases, these statements are made without any formal attribution studies, and in some cases subsequent attribution research has shown that public statements on probable causes are extremely unlikely (Hoerling *et al.*, 2007).

The ability to attribute observed climate variations and change provides an essential component within a comprehensive climate information system designed to serve a broad range of public needs (Trenberth *et al.*, 2006; NIDIS, 2007). Reliable attribution provides a scientific underpinning for improving climate predictions and climate change projections and information useful for evaluating policy options and responses and managing resources. This capability is also vital for assessing climate model performance and identifying where future model improvements are most needed. The associated scientific capacity should include providing coordination of and access to critical observational and reanalysis datasets as well as output from model experiments in which different forcings are systematically included or excluded. Without a clear and systematic institutional commitment, future efforts in climate attribution are likely to continue to be *ad hoc*, and unlikely to be conducted as efficiently and effectively as possible.

Reliable attribution provides a scientific underpinning for improving climate predictions and climate change projections and information useful for evaluating policy options and responses and managing resources.

In order to develop this capacity, there is a need to improve coordination of, and access to, climate model and observational data relevant for climate attribution. Compared with earlier climate change assessments, a major advance in the Intergovernmental Panel on Climate Change (IPCC) Fourth Assessment Report was the much larger number of simulations obtained from a broader range of models (IPCC, 2007). Taken together with additional observations, these more extensive simulations helped provide for the first time quantitative estimates of the likelihoods of certain aspects of future climate change. This work was facilitated substantially through the Program for Climate Model Diagnosis and Intercomparison (PCMDI), which provided facilities for storing and distributing the very large datasets that were generated from the numerous climate model simulations of past climate and climate change projections that were generated for the IPCC report. Other basic infrastructure tasks provided through PCMDI included: the development of software for data management; visualization and computation; the assembly and organization of observational datasets for model validation; and consistent documentation of climate model features. Providing similar infrastructure support for a broader range of necessary model simulations will be vital to continuing advances in research on climate attribution. In addition to fundamental data management responsibilities, advances in scientific visualization and diagnostic and statistical methods for intercomparing and evaluating results from model simulations would substantially facilitate future research.

The continual interplay between observations and models for climate analysis and reanalysis that occurs in attribution studies is fundamental to achieving the long-term objectives of the CCSP (CCSP, 2003). Detection and attribution research provide a rigorous comparison between model-simulated and observed climate changes. Climate variations and changes that can be detected and attributed to factors external to the climate system, such as from solar variations, greenhouse gas increases produced by human activities, or aerosols ejected into the atmosphere from volcanic eruptions, help to constrain uncertainties in future predictions and projections of climate variations and change.

Climate variations that can be attributed to factors that are internal to the climate system, such as sea surface temperature or soil moisture conditions, can also help constrain uncertainties in future predictions of climate variations over time periods of seasons to decades. At the same time, where there are significant discrepancies between model simulations and observations that are outside the range of natural climate variability, the information provided through detection and attribution studies helps identify model deficiencies and areas where additional effort will be required to reduce uncertainties in climate predictions and climate change projections.

- **Focus research to better explain causes of climate conditions at regional and local levels, including the roles of changes in land cover, land use, atmospheric aerosols, greenhouse gases, sea surface temperatures, and other factors that contribute to climate change.**

While significant advances have occurred over the past decade in attributing causes for observed climate variations and change, there remain important sources for uncertainties. These sources become increasingly important in going from global to regional and local scales. They include: uncertainties in observed magnitudes and distributions of forcing from various processes; uncertainties in responses to various forcings; uncertainties in natural variability in the climate system, that is, variability that would occur even in the absence of changes in external forcing.

To address these uncertainties, further research is needed to improve observational estimates of changes in radiative forcing factors over a reference time period, for example, the twentieth century to the present. In addition to greenhouse gas changes, such factors include variations in solar forcing, effects of atmospheric aerosols, and land use and land cover changes. The relative importance of these factors varies among climate variables, and space and time scales. For instance, land use changes are likely to have a relatively small effect in changing global average temperature (*e.g.*, Matthews *et al.*, 2004) but may have more substantial effects on local weather (*e.g.*, Pielke *et al.*, 1999; Chase *et al.*,

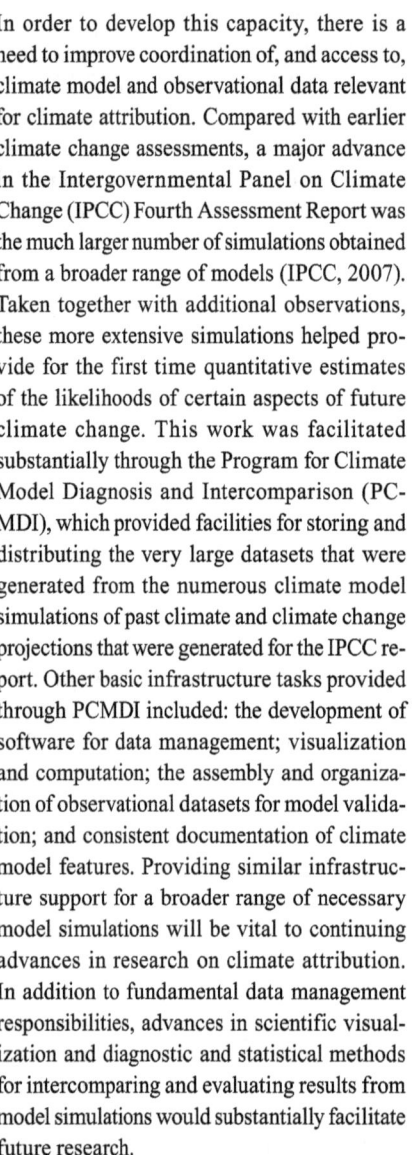

There is a need to improve coordination of, and access to, climate model and observational data relevant for climate attribution.

2000; Baidya and Avissar, 2002; Pielke, 2001). Aerosol variations are also likely to be increasingly important in forcing climate variations at regional to local levels (Kunkel *et al.*, 2006). Detection and attribution results are sensitive to forcing uncertainties, which can be seen when results from models are compared with different forcing assumptions (*e.g.*, Santer *et al.*, 1996; Hegerl *et al.*, 2000; Allen *et al.*, 2006).

More comprehensive and systematic investigations are also required of the climate response to individual forcing factors, as well as to combinations of factors. Parallel efforts are necessary to estimate the range of unforced natural variability and model climate drift. Ensemble model experiments should be performed with a diverse set of coupled climate models over a common reference period, such as the twentieth century to present, in which different factors are systematically included or excluded. For example, model simulations including and excluding changes in observed land cover are needed to better quantify the potential influence of anthropogenic land cover change, especially at regional or smaller levels. Extended control simulations are required with the same models to estimate natural internal variability and assess model climate drifts. The ability to carry out the extensive simulations that are required will depend strongly on the availability of high performance computing capabilities.

A first estimate of combined model errors and forcing uncertainties can be determined by combining data from simulations forced with different estimates of radiative forcings and simulated with different models (Hegerl *et al.*, 2006). Such multi-model fingerprints provide an increased level of confidence in attribution of observed warming from increases in greenhouse gases and cooling from sulfate aerosols (Gillett *et al.*, 2002). Both forcing and model uncertainties need to be explored more completely in order to better understand the effects of forcing and model uncertainty, and their representation in detection and attribution (Hasselmann, 1997). Because the use of a single model may lead to underestimates of the true uncertainty, it is important that such experiments reflect a diversity of responses as obtained from a broad range of models (Hegerl *et al.*, 2006).

As discussed in Chapter 3, atmospheric models forced by observed changes in sea-surface temperatures have shown considerable ability to reproduce aspects of climate variability and change over North America and surrounding regions since 1950. A growing body of evidence indicates that changes in the oceans are central to understanding the causes of other major climate anomalies. Additional assessments are required to better determine the atmospheric response to sea surface temperature variations and, in particular, the extent to which changing ocean conditions may account for past and ongoing climate variations and change. As part of this assessment, ensemble experiments should be conducted with atmospheric models forced by observed sea-surface temperatures over the same baseline time period and in parallel with the experiments recommended earlier.

- **Explore a range of methods to better quantify and communicate findings from attribution research.**

There is a need to develop alternative approaches to more effectively communicate knowledge on the causes of observed climate variability and change, and potential implications for decision makers (*e.g.*, for risk assessment). New methods will become increasingly important in considering variability and changes at smaller space and time scales than in traditional global change studies, as well as for assessments of factors contributing to the likelihood of extreme weather and climate events. There is a strong need to go beyond present communication methods to approaches that include specific

More comprehensive and systematic investigations are required of the climate response to individual forcing factors, as well as to combinations of factors. Parallel efforts are necessary to estimate the range of unforced natural variability and model climate drift.

responsibilities for addressing questions of public interest.

Much of the climate attribution research to date has focused on identifying the causes for long-term climate trends. An important new challenge is quantifying the impact of various factors that influence the probability of specific weather or short-term climate events (CCSP, 2008). An often-stated assertion is that it is impossible to attribute a single event in a chaotic system to external forcing, although it is through such events that society experiences many of the impacts of climate variability and change. As discussed in Hegerl *et al.* (2006), this statement is based in part on an underlying statistical model that assumes that what is observed at any time is a deterministic response to forcing upon which is superposed random "climate noise". From such a model, it is possible to estimate underlying deterministic changes in certain statistical properties, for example, expected changes in event frequency over time, but not to attribute causes for individual events themselves.

In order to be more responsive to questions from government, media, and the public, a coordinated, ongoing activity in climate attribution should include specific responsibilities for addressing questions of public and private interests on the causes of observed climate variations and change.

However, several recent studies demonstrate that quantitative probabilistic attribution statements are possible for individual weather and climate events, if the statements are framed in terms of the contribution of the external forcing to changes in the relative likelihood of occurrence of the event (Allen, 2003; Stone and Allen, 2005; Stott *et al.*, 2004). Changes in likelihood in response to a forcing can be stated in terms of the "fraction of attributable risk" (FAR) due to that forcing. The FAR has a long-established use in fields such as epidemiology, for example, in determining the contribution of a given risk factor (*e.g.*, tobacco smoking) to disease occurrence (*e.g.*, lung cancer). This approach has been applied to attribute a fraction of the probability of an extreme heat wave observed in Europe in 2003 to anthropogenic forcing (Stott *et al.*, 2004) and more recently, to the extreme annual U.S. warmth of 2006 (Hoerling *et al.*, 2007). These probabilistic attribution findings related to risk assessment should be explored further, as this information may be more readily interpretable and usable by many decision makers.

There is also a strong need to go beyond present limited efforts at communicating knowledge on the causes of observed climate variations and change. In order to be more responsive to questions from government, media, and the public, a coordinated, ongoing activity in climate attribution should include specific responsibilities for addressing questions of public and private interests on the causes of observed climate variations and change. This capability will form a necessary collaborative component within a national climate information system designed to meet the core CCSP objective of providing science-based information for improved decision support (CCSP, 2003).

APPENDIX A

Data Assimilation

Coordinating Lead Authors: James Carton, University of Maryland; Eugenia Kalnay, University of Maryland

Data assimilation is an exercise in the calculation of conditional probabilities in which short-term model forecasts are combined with observations to best estimate the state of, for example, the atmosphere. Since there are limitations in model resolution and errors associated with parameterization of unresolved physical processes, and the behavior of the atmosphere is chaotic, forecast accuracy is described by a probability distribution, as is observation accuracy. These probability distributions are combined to form conditional probabilities, which are simplified by assuming these distributions are Gaussian (normally distributed). The conditional probabilities are used to create a more accurate *analysis* than can be obtained solely from either the forecasts or the observations. The same approach can be applied to the ocean, land surface, or cryosphere.

Atmospheric data assimilation proceeds through a succession of (typically) six hour *analysis cycles*. At the beginning of each cycle, a six-hour model forecast is carried out starting from initial conditions of atmospheric pressure, temperature, humidity, and winds provided by the previous analysis cycle, with observed boundary conditions such as sea surface temperature and snow cover. At the end of each cycle all available current observations are quality controlled, and the differences between the observations and the model forecast of the same variables, referred to as observational increments or innovations, are computed. The observations may include the same variables observed with different systems (*e.g.*, winds measured from airplanes or by following the movement of clouds). They may also include observations of variables that do not directly enter the forecast such as satellite radiances, which contain information about both temperature and moisture.

If the evolving probability distributions of the model forecasts and observations are known, then it is possible to construct an analysis that is optimal because the expected error variance, which is the difference between the analysis of a variable and its true value, is minimized. In practice, the probability distributions are unknown. In addition, it is not possible to solve the computational problem of minimizing the error variance for realistic complex systems. In order to address these problems, several simplifying assumptions are needed. The observational increments are generally assumed to be Gaussian. With this assumption a cost function can be constructed whose minimization, which provides us with the optimal analysis, leads to the Kalman Filter equations. A bigger assumption that the probability distribution of the forecast errors does not depend on time, gives rise to the widely used and more simplistic three-dimensional variational type of data assimilation (3DVAR). Four-dimensional variational data assimilation (4DVAR) is a generalization of the cost function approach that allows the forecast initial conditions (or other control variables such as diffusive parameters) to be modified based on observations within a time window.

Despite the use of simplifying assumptions, the Kalman Filter and 4DVAR approaches still lead to challenging computational problems. Efforts to reduce the magnitude of the computational problems and exploit physical understanding of the physical system have led to the development of Monte Carlo approaches known as Ensemble Kalman Filter (EnKF). EnKF methods, like 4DVAR, can be posed in such a way that the analysis at a given time can be influenced by past, present, and future observations. This property of time symmetry is especially desirable in reanalyses since it allows the analysis at past times to benefit to some extent from future enhancements of the observing system.

Table A.1 Characteristics of some existing global ocean model-based reanalyses of ocean climate (extracted from: <http://www.clivar.org/data/synthesis/directory.php>)

Organization/System	Model	Analysis Method	Time Period	Weblinks
NCEP, Météo France, CERFACS	OPA8.2,2°x2°x31 Lev (~0.5°x2° tropics) ERA40 forcing	Multivariate 3D-Var (OPAVAR) for T & S profiles	1962 to 2001	<.fr/globc/overview.html>
ECMWF	HOPE, 1°x1°x29 Lev (1/3°x1° tropics)	OI	1959 to 2006	<ecmwf.int/products/forecasts/d/charts/ocean/reanalysis/>
ECCO-GODAE	MITgcm 1°x1°	4DVAR	1992 to 2004	<www.ecco-group.org>
ECCO-JPL	MITgcm and MOM4 1°x1°x50 lev	Kalman filter and RTS smoother	1993 to present	<ecco.jpl.nasa.gov/external/>
ECCO-SIO	1°x1°	4DVAR	1992 to 2002	<ecco.ucsd.edu>
ECCO2	MITgcm, 18kmx50 Lev	Green's functions	1992 to present	
ENACT consortium			1962 to 2006	<www.ecmwf.int/research/EU_projects/ENACT/>
FNMOC/GODAE				<www.usgodae.org>
GECCO			1950 to 2000	<www.ecco-group.org>
GFDL			1960 to 2006	<www.gfdl.noaa.gov/>
UK Met Office GloSea	GloSea OGCM 1.25°x1.25°x40 Lev (0.3°x1.25° tropics) daily ERA40 fluxes with corrected precipitation	OI	1962 to 1998	<www.metoffice.gov.uk/research/seasonal/glosea.html>
NASA Goddard GMAO	Poseidon, 1/3°x5/8°	MVOI, Ensemble KF	1993 to present	<gmao.gsfc.nasa.gov>
INGV	OPA8.2 2°x2°x31 Lev (0.5°x2° tropics) ERA40 and operational ECMWF fluxes	Reduced Order MVOI with bivariate T and S EOFs	1962 to present	
MEXT K-7	MOMv3 1°x1°x36 Lev NCEP2 reanalysis, ISCCP data	4DVAR	1990 to 2000	<www.jamstec.go.jp/frcgc/k7-dbase2/eng/>
MERCATOR-3	OPA8.2 2°x2°x31 Lev (~0.5° meridional at the tropics)	Single Evolutive Extended Kalman (SEEK) filter	1993 to 2001	<www.mercator-ocean.fr.html/systemes_ops/psy3/index_en.html>
JMA MOVE/MRI.COM			1949 to 2005	<www.mri-jma.go.jp/Dep/oc/oc.html>
NOAA/NCEP GODAS	MOMv3 1°x1°x40 Lev (1/3°x1° tropics) NCEP Reanalysis2	3DVAR	1980 to present	<www.bom.gov.au/bmrc/ocean/JAFOOS/POAMA/>
BoM, CSIRO, POAMA	ACOM2 (based on MOM2), 2°x2°x27 Lev (0.5°x2° at high latitudes) ERA40	MVOI, ensemble KF	1980 to 2006	<www.atmos.umd.edu/~ocean/>
SODA	POP1.4, POP2.01, global ave	MVOI with evolving error	1958 to 2005	

APPENDIX B

Data and Methods Used for Attribution

Convening Lead Author: Martin Hoerling, NOAA

Lead Authors: Gabriele Hegerl, University of Edinburgh; David Karoly, University of Melbourne; Arun Kumar, NOAA; David Rind, NASA

B.1 OBSERVATIONAL DATA

North American surface temperatures during the assessment period of 1951 to 2006 are derived from four data sources, which include: the U.K. Hadley Centre's HadCRUT3v (Brohan *et al.*, 2006); NOAA's land/ocean merged data (Smith and Reynolds, 2005); NOAA's global land gridded data (Peterson *et al.*, 1998); and NASA's gridded data (Hansen *et al.*, 2001). For analysis of U.S. surface temperatures, two additional datasets used are NOAA's U.S. Climate Division data (NCDC, 1994) and PRISM data.

Spatial maps of the surface temperature trends shown in Chapter 3 are based on combining all the above datasets. For example, the North American and U.S. surface temperature trends during 1951 to 2006 were computed for each dataset, and the trend map is based on equal-weighted averages of the individual trends. The uncertainty in observations is displayed by plotting the extreme range among the time series of the 1951 to 2006 trends from individual datasets.

North American precipitation data are derived from the Global Precipitation Climatology Project (GPCC) (Rudolf and Schneider, 2005); the NOAA gridded precipitation data has also been consulted (Chen *et al.*, 2002). However, the North American analysis shown in Chapter 3 is based on the GPCC data alone which is judged to be superior, owing to its greater volume of input stations over Canada and Alaska in particular. For analysis of U.S. precipitation, two additional datasets used are NOAA's U.S. Climate Division data and PRISM data. Spatial maps of U.S. precipitation trends during 1951 to 2006 were computed for each of these three datasets, and the U.S. trend map is based on equal-weighted averages of the individual trends.

Free atmospheric conditions during 1951 to 2006, including 500 hPa geopotential heights, are derived from the NCEP/NCAR reanalysis (Kalnay *et al.*, 1996). A comparison of various reanalysis data is provided in Chapter 2, but only the NCEP/NCAR version is available for the entire 1951 to 2006 assessment period.

B.2 CLIMATE MODEL SIMULATION DATA

Two configurations of climate models are used in this Report: atmospheric general circulation models (AMIP), and coupled ocean-atmosphere general circulation models (CMIP). For the former, the data from two different atmospheric models are studied; the European Center/ Hamburg model (ECHAM4.5) (Roeckner *et al.*, 1996) whose simulations were performed by the International Research Institute for Climate and Society at LaMont Doherty (L. Goddard, personal communication), and the NASA Seasonal-to-Interannual Prediction Project (NSIPP) model (Schubert *et al.*, 2004) whose simulations were conducted at NASA/Goddard. The models were subjected to specified monthly varying observed global sea surface temperatures during 1951 to 2006. In a procedure that is commonly used in climate science, multiple realizations of the 1951 to 2006 period were conducted with each model in which the separate runs started from different atmospheric initial conditions but were subjected to identically evolving SST conditions. A total of 33 AMIP runs (24 ECHAM and 9 NASA) were available.

The coupled models are those used in the IPCC Fourth Assessment Report (IPCC, 2007a). These are forced with estimated greenhouse gases, atmospheric aerosols, solar irradiance, and the radiative effects of volcanic activity for 1951 to 1999, and with the IPCC Special Emissions Scenario (SRES) A1B (IPCC, 2007a) for 2000 to 2006. The model data are available from the Program for Climate Model Diagnosis and Intercomparison (PCMDI) archive as part of the Coupled Model Intercomparison Project (CMIP3). Table 3.1 lists the 19 different models used and the number of realizations conducted with each model. A total of 41 runs were available.

The SST-forced (externally-forced) signal of North American and U.S. surface temperature and precipitation variability during 1951 to 2006 is estimated by averaging the total of 33 AMIP (41 CMIP) simulations. Trends during 1951 to 2006 were computed for each model run in a manner identical to the observational method; the trend map shown in Chapter 3 is based on an equal-weighted ensemble average of the individual trends. The uncertainty in these simulated trends is displayed graphically by plotting the 5 to 95 percent range amongst the individual model runs.

All the observational and model data used in this Product are available in the public domain (see Table 3.2 for website information). Further, these data have been widely used for a variety of climate analysis studies as reported in the refereed scientific literature.

B.3 DATA ANALYSIS AND ASSESSMENT

Analysis of observational and model data is based on standard statistical procedures used extensively in climate research and the physical sciences (von Storch and Zwiers, 1999). Trends for 1951 to 2006 are computed using a linear methodology based on least squares which is a mathematical method of finding a best fitting curve by minimizing the sums of the squares of the residuals. Statistical estimates of the significance of the observed trends are based on a non-parametric test in which the 56-year trends are ranked against those computed from CMIP simulations subjected to only natural forcing (solar irradiance and volcanic aerosol). The principal uncertainty in such an analysis is knowing the population (number) of 56-year trends that are expected in the absence of anthropogenic forcing. Chapter 3 uses four different coupled models, and a total of sixteen 100-year simulations to estimate the statistical population of naturally occurring 56-year trends, though the existence of model biases is taken into account in making expert assessments.

Observed and modeled data are compared using routine linear statistical methods. Time series are intercompared using standard temporal correlations. Spatial maps of observed and simulated trends over North America are compared using standard spatial correlation and congruence calculations. Similar empirical methods have been applied for pattern analysis of climate change signals in the published literature (Santer *et al.*, 1994).

Expert judgment is used in Chapter 3 to arrive at probabilistic attribution statements. The analyses described above are only a small part of the information available to the authors, who also make extensive use of the scientific peer-reviewed literature. For more details on the use of expert assessment in this Product, the reader is referred to Box 3.4 and the Preface.

GLOSSARY AND ACRONYMS

GLOSSARY

This glossary defines some specific terms within the context of this Product. Most terms below are adapted directly from definitions provided in the Intergovernmental Panel on Climate Change (IPCC) Fourth Assessment Report Glossary. Those terms not included in the IPCC report or whose definitions are not identical to the usage in the IPCC Glossary are marked with an asterisk.

abrupt climate change
The non-linearity of the climate system may lead to abrupt climate change, sometimes called *rapid climate change, abrupt events* or even *surprises*. The term "abrupt" often refers to changes that occur on time scales faster than the typical time scale of the responsible forcing. However, abrupt climate changes need not be externally forced, and rapid transitions can result simply from physical or dynamical processes internal to the climate system.

aerosols
A collection of airborne solid or liquid particles, with a typical size between 0.01 and 10 micrometers (μm) and residing in the atmosphere for at least several hours. Aerosols may be of either natural or anthropogenic origin.

analysis*
A detailed representation of the state of the atmosphere and, more generally, other components of the climate system, such as oceans or land surfaces, that is based on observations.

annular modes
Preferred patterns of change in atmospheric circulation corresponding to changes in the zonally averaged midlatitude westerlies. The Northern Annular Mode has a bias to the North Atlantic and has a large correlation with the North Atlantic Oscillation. The Southern Annular Mode occurs in the Southern hemisphere.

anthropogenic
Resulting from or produced by the activities of human beings.

attribution*
The process of establishing the most likely causes for a detected climate variation or change with some defined level of confidence.

climate
The statistical description in terms of the mean and variability of relevant atmospheric variables over a period of time ranging from months out to decades, centuries, and beyond. Climate conditions are often described in terms of surface variables such as temperature, precipitation, and wind. Climate in a wider sense is a description of the full climate system, including: the atmosphere, the oceans, the cryosphere, the land surface, and the biosphere, as well as their interactions.

climate change
A change in the state of the climate that can be identified (*e.g.*, using statistical tests) by changes in the mean and/or the variability of its properties, and that persists for an extended period, typically decades or longer. Climate change may be due to natural internal processes or external forcings, or to persistent anthropogenic changes in the composition of the atmosphere or in land use.

climate system
The climate system is the highly complex system consisting of five major components: the atmosphere, the hydrosphere, the cryosphere, the land surface and the biosphere, and the interactions between them. The climate system evolves in time under the influence of its own internal dynamics and because of external forcings such as volcanic eruptions, solar variations and human-induced forcings such as the changing composition of the atmosphere and changes in land cover and land use.

climate variability
Variations in the mean state and other statistics (such as standard deviations, the occurrence of extremes, *etc.*) of the climate on all temporal and spatial scales beyond that of individual weather events. Variability may be due to natural internal processes within the *climate system* (internal variability), or to variations in natural or anthropogenic external forcing (external variability).

confidence
The likelihood of the correctness of a result as expressed in this Product, using a standard terminology defined in the Preface.

data assimilation*
The combining of diverse observations, possibly sampled at different times and intervals and different locations, into a unified and consistent description of a physical system, such as the state of the atmosphere. This combination is obtained by integrating the observations together in a

109

numerical prediction model that provides an initial estimate of the state of the system, or "first guess".

drought

In general terms, drought is a "prolonged absence or marked deficiency of precipitation", a "deficiency that results in water shortage for some activity or for some group", or a "period of abnormally dry weather sufficiently prolonged for the lack of precipitation to cause a serious hydrological imbalance" (Heim, 2002). Related terms include the following: *Agricultural drought* relates to moisture deficits in the topmost meter or so of soil (the root zone) that impacts crops, *meteorological drought* is mainly a prolonged deficit of precipitation, and *hydrologic drought* is related to below normal streamflow, lake and groundwater levels. A *megadrought* is a long, drawn-out, and pervasive drought, lasting much longer than normal, usually a decade or more.

El Niño-Southern Oscillation (ENSO)

El Niño, in its original sense, is a warm water current that periodically flows along the coast of Ecuador and Perú, disrupting the local fishery. It has since become identified with a basin-wide warming of the tropical Pacific east of the dateline. This oceanic event is associated with a fluctuation of a global scale tropical and subtropical surface pressure pattern, called the Southern Oscillation. This coupled atmosphere-ocean phenomenon, with preferred time scales of two to about seven years, is collectively known as El Niño-Southern Oscillation, or ENSO. ENSO is often measured by the surface pressure anomaly difference between Darwin, Australia and Tahiti, and the sea surface temperatures in the central and eastern equatorial Pacific. During an ENSO event the prevailing trade winds weaken, reducing upwelling and altering ocean currents such that the sea surface temperatures warm, further weakening the trade winds. This event has great impact on the wind, sea surface temperature and precipitation patterns in the tropical Pacific. It has climatic effects throughout the Pacific region and in many other parts of the world, through global teleconnections with fluctuations elsewhere. The cold phase of ENSO is called *La Niña*.

ensemble

A group of parallel model simulations. Typical ensemble sizes in many studies range from 10 to 100 members, although this number is often considerably smaller for long runs with the most complex climate models. Variation of the results across the ensemble members gives an estimate of uncertainty. Ensembles made with the same model but different initial conditions characterize the uncertainty associated with internal climate variability, whereas multi-model ensembles including simulations by several models also include effects of model differences. Perturbed-parameter ensembles, in which model parameters are varied in a sys-

tematic manner, aim to produce a more objective estimate of modeling uncertainty than is possible with traditional multi-model ensembles.

evapotranspiration

The combined process of evaporation from the Earth's surface and transpiration from vegetation.

fingerprint

The climate response pattern in space and/or time to a specific forcing. Fingerprints are used to detect the presence of this response in observations and are typically estimated using forced climate model simulations.

geostrophic wind (or current)

A wind or current that represents a balance between the horizontal pressure gradient and the Coriolis force. The geostrophic wind or current flows directly parallel to isobars with a speed inversely proportional to the spacing of the isobaric contours (*i.e.*, tighter spacing implies stronger geostrophic winds). This is one example of an important balance relationship between two fundamental fields, mass (represented by pressure) and momentum (represented by winds), and implies that information about one of those two fields also implies information on the other.

land use and land-use change

Land use refers to the total of arrangements, activities and inputs undertaken in a certain land cover type (a set of human actions). The term "land use" is also used in the sense of the social and economic purposes for which land is managed (*e.g.*, grazing, timber extraction, and conservation).
Land-use change refers to a change in the use or management of land by humans, which may lead to a change in land cover. Land cover and land-use change may have an impact on the surface albedo, evapotranspiration, sources and sinks of greenhouse gases, or other properties of the climate system and may thus have a radiative forcing and/or other impacts on climate, locally or globally.

likelihood

The probability of an occurrence, an outcome or a result. This is expressed in this Product using a standard terminology, as defined in the Preface.

modes of climate variability

Natural variability of the climate system, in particular on seasonal and longer timescales, predominantly occurs with preferred spatial patterns and timescales, through the dynamical characteristics of the atmospheric circulation and through interactions with the land and ocean surfaces. Such patterns are often called *regimes* or *modes* or Pacific North American pattern (PNA), the El Niño-Southern Oscillation (ENSO), the Northern Annular Mode (NAM; previously called Arctic Oscillation, AO) and the Southern Annular Mode (SAM; previously called Antarctic Oscillation, AAO). Many of the prominent modes of climate variability are discussed in Chapter 2.

non-linearity
A process where there is no simple proportional relation between cause and effect. The climate system contains many such non-linear processes, resulting in a system with a potentially very complex behavior. Such complexity may lead to abrupt climate change.

North Atlantic Oscillation (NAO)
The North Atlantic Oscillation is defined by opposing variations of barometric pressure near Iceland and near the Azores. Through the geostrophic wind relationship, it also corresponds to fluctuations in the strength of the main westerly winds across the Atlantic into Europe, and thus also influences storm tracks that influence these regions.

Northern Annular Mode (NAM)
A winter-time fluctuation in the amplitude of a pattern characterized by low surface pressure in the Arctic and strong middle latitude westerlies. The NAM has links with the northern polar vortex into the stratosphere. Its pattern has a bias to the North Atlantic and has a large correlation with the North Atlantic Oscillation.

numerical prediction model*
A model that predicts the evolution of the atmosphere (and more generally, other components of the climate system, such as the ocean) through numerical methods that represent the governing physical and dynamical equations for the system. Such approaches are fundamental to almost all dynamical weather prediction schemes, since the complexity of the governing equations do not allow exact solutions.

Pacific Decadal Variability
Coupled decadal-to-interdecadal variability of the atmospheric circulation and underlying ocean in the Pacific basin. It is most prominent in the North Pacific, where fluctuations in the strength of the wintertime Aleutian Low pressure system co-vary with North Pacific sea surface temperature, and are linked to decadal variations in atmospheric circulation, sea surface temperature and ocean circulation throughout the whole Pacific Basin.

Pacific North American (PNA) pattern
An atmospheric large-scale wave pattern featuring a sequence of tropospheric high and low pressure anomalies stretching from the subtropical west Pacific to the east coast of North America.

paleoclimate
Climate during periods prior to the development of measuring instruments, including historic and geologic time, for which only proxy climate records are available.

parameterization
The technique of representing processes that cannot be explicitly resolved at the spatial or temporal resolution of the model (sub-grid scale processes), by relationships between model-resolved larger scale flow and the area or time averaged effect of such sub-grid scale processes.

patterns of climate variability
Natural variability of the climate system, in particular on seasonal and longer time-scales, predominantly occurs with preferred spatial patterns and timescales, through the dynamical characteristics of the atmospheric circulation and through interactions with the land and ocean surfaces. Such patterns are often called regimes, modes or teleconnections. Examples are the North Atlantic Oscillation (NAO), the Pacific-North American pattern (PNA), the El Niño-Southern Oscillation (ENSO), and the Northern and Southern Annual Mode (NAM and SAM). Many of the prominent modes of climate variability are discussed in Chapter 2.

predictability
The extent to which future states of a system may be predicted based on knowledge of current and past states of the system.

probability density function (PDF)
A probability density function is a function that indicates the relative chances of occurrence of different outcomes of a variable.

reanalysis*
An objective, quantitative method for representing past weather and climate conditions and, more generally, conditions of other components of the Earth's climate system such as the oceans or land surface. An important goal of most reanalysis efforts to date has been to reconstruct a detailed, accurate, and continuous record of past global atmospheric conditions, typically at time intervals of every 6 to 12 hours, over periods of decades or longer. This reconstruction is accomplished by integrating observations obtained from numerous data sources together within a numerical prediction model through a process called data assimilation.

sea surface temperature
The bulk temperature in the top few meters of the ocean. Measurements are made by ships, buoys and drifters.

storm tracks
Originally a term referring to the tracks of individual cyclonic weather systems, but now often generalized to refer to the regions where the main tracks of extratropical disturbances occur as sequences of low (cyclonic) and high (anticyclonic) pressure systems.

stratosphere
The highly stratified region of the atmosphere above the troposphere extending from about 10 kilometers (km) (ranging from 9 km in high latitudes to 16 km in the tropics on average) to about 50 km altitude.

teleconnection
A connection between climate variations over widely separated parts of the world. In physical terms, teleconnections are often a consequence of large-scale wave motions, whereby energy is dispersed from source regions along preferred paths in the atmosphere.

troposphere
The lowest part of the atmosphere from the surface to about 10 kilometers (km) in altitude in midlatitudes (ranging from 9 km in high latitudes to 16 km in the tropics on average) where clouds and weather phenomena occur. In the troposphere temperatures generally decrease with height.

ACRONYMS

AGCM	Atmospheric General Circulation Model
AMIP	Atmospheric Model Intercomparison Project
AMO	Atlantic Multi-decadal Oscillation
AMS	American Meteorological Society
AR4	IPCC Fourth Assessment Report
BC	black carbon
CCCma-CGCM3.1(T47)	a Canadian Centre for Climate Modelling and Analysis model
CCSM3	a National Center for Atmospheric Research model
CCSP	Climate Change Science Program
CFS	Climate Forecast System
CFSRR	Climate Forecast System Reanalysis and Reforecast Project
CMIP	Coupled Model Intercomparison Project
CNRM-CM3	aMétéo-France/Centre National de Recherches Météorologiques model
CRU	Climate Research Unit
CRUTEM	Climate Research Unit Land Temperature Record
CSIRO	Commonwealth Scientific and Industrial Organization
CSIRO-Mk3.0	a CSIRO Marine and Atmospheric Research model
CTD	Conductivity Temperature Depth
DJF	December-January-February
DOE	Department of Energy
ECHAM5/MPI-OM	a Max-Planck Institute for Meteorology model
ECMWF	European Center for Medium-Range Weather Forecasting
ENSO	El Niño-Southern Oscillation
ESMF	Earth System Modeling Framework
EU	European Union
FAR	fraction of attributable risk
FGGE	First GARP Global Experiment
FGOALS-g1.0	an Institute for Atmospheric Physics model
GARP	GEMPAK Analysis and Rendering Program
GCHN	Global Historical Climatology Network
GCM	Global Circulation Model
GCOS	Global Climate Observing System
GEMPAK	General Meteorology Package
GEMS	Global Environment Monitoring System
GEOS	Goddard Earth Observing System
GEOSS	Global Earth Observing System of Systems

GFDL	Geophysical Fluid Dynamics Laboratory	**NAO**	North Atlantic Oscillation
GFDL-CM2.0	a Geophysical Fluid Dynamics Laboratory model	**NARR**	North American Regional Reanalysis
GFDL-CM2.1	a Geophysical Fluid Dynamics Laboratory model	**NASA**	National Aeronautics and Space Administration
GISS	Goddard Institute for Space Studies	**NCAR**	National Center for Atmospheric Research
GISS-EH	a Goddard Institute for Space Studies model	**NCDC**	National Climatic Data Center
GISS-ER	a Goddard Institute for Space Studies model	**NCEP**	National Centers for Environmental Prediction
GMAO	Global Modeling and Assimilation Office	**NIDIS**	National Integrated Drought Information System
GODAR ogy	Global Oceanographic Data Archaeol- and Rescue	**NIES**	National Institute for Environmental Studies
GPCC	Global Precipitation Climatology Project	**NOAA**	National Oceanic and Atmospheric Administration
GRIPS	GCM-Reality Intercomparison Project for SPARC	**NRC**	National Research Council
GSI	grid-point statistical interpolation	**NSIPP**	NASA Seasonal-to-Interannual Prediction Project
HIRS	High-resolution Infrared Radiation Sounder	**OSE**	Observing System Experiments
ICOADS	International Comprehensive Ocean-Atmosphere Data Set	**PCM**	National Center for Atmospheric Research model
IDAG	International Ad Hoc Detection and Attribution Group	**PCMDI**	Program for Climate Model Diagnosis and Intercomparison
IESA	integrated Earth system analysis	**PDO**	Pacific Decadal Oscillation
INM-CM3.0	an Institute for Numerical Mathematics model	**PDSI**	Palmer Drought Severity Index
IPCC	Intergovernmental Panel on Climate Change	**PIRATA**	Pilot Research Moored Array in the Atlantic
IPSL-CM4	Institute Pierre Simon Laplace model	**PNA**	Pacific North American Pattern
ITCZ	Intertropical Convergence Zone	**PRISM**	Precipitation-elevation Regressions on Independent Slopes Model
JAMSTEC	Frontier Research Center for Global Change in Japan	**QBO**	Quasi-Biennial Oscillation
JJA	June-July-August	**SAP**	Synthesis and Assessment Product
LDAS	Land Data Assimilation System	**SNOTEL**	Snowpack Telemetry
LLJ	low-level jet	**SODA**	Simple Ocean Data Assimilation
MERRA	Modern Era Retrospective-Analysis for Research and Applications	**SPARC**	Stratospheric Processes and their Role in Climate
MIROC3.2(medres)	a Center for Climate System Research model	**SRES**	(IPCC) Special Emissions Scenario
		SST	sea surface temperature
MIROC3.2(hires)	a Center for Climate System Research model	**SSU**	Stratospheric Sounding Unit
		TAO	Tropical Atmosphere Ocean
MJO	Madden-Julian Oscillation	**TAR**	IPCC Third Assessment Report
MRI	Meteorological Research Institute	T_{2m}	two meter height temperature
MRI-CGCM2.3.2	a Meteorological Research Institute model	**UKMO-HadCM3**	a Hadley Centre for Climate Prediction and Research model
MSU	Microwave Sounding Unit	**UKMO-HadGEM1**	a Hadley Centre for Climate Prediction and Research model
NAM	Northern Annular Mode	**WCRP**	World Climate Research Programme
NAMS	North American Monsoon System	**WOAP**	WCRP Observations and Assimilation Panel
		WOD	World Ocean Database
		XBT	expendable bathythermograph

REFERENCES

CHAPTER 1 REFERENCES

Challinor, A.J., T.R. Wheeler, J.M. Slingo, P.Q. Crauford, and D. Grimes, 2005: Simulation of crop yields using ERA-40: Limits to skill and nonstationarity in weather-yield relationships. *Journal of Applied Meteorology*, **44(4)**, 516-531.

Fuchs, T., 2007: *GPCC Annual Report for the year 2006: Development of the GPCC Data Base and Analysis Products*. Deutcher Wetterdienst, Offenbach, (Germany), 14 pp. <http://www.dwd.de/bvbw/appmanager/bvbw/dwd-wwwDesktop?_nfpb=true&_pageLabel=dwdwww_result_page&gsbSearchDocId=190876>

Geer, I.W. (ed.), 1996: *Glossary of Weather and Climate, with Related Oceanic and Hydrologic Terms*. American Meteorological Society, Boston, MA, 272 pp.

IPCC (Intergovernmental Panel on Climate Change), 2007a: *Climate Change 2007: The Physical Science Basis*. Contribution of Working Group I to the Fourth Assessment Report (AR4) of the Intergovernmental Panel on Climate Change [Solomon, S., D. Qin, M. Manning, Z. Chen, M. Marquis, K.B. Averyt, M. Tignor, and H.L. Miller (eds.)]. Cambridge University Press, Cambridge, UK, and New York, 996 pp. <http://www.ipcc.ch>

IPCC (Intergovernmental Panel on Climate Change), 2007b: *Climate Change 2007: Impacts, Adaptation and Vulnerability*. Contribution of Working Group II to the Fourth Assessment Report (AR4) of the Intergovernmental Panel on Climate Change [Parry, M.L., O.F. Canziani, J.P. Palutikof, P.J. van der Linden, and C.E. Hanson (eds.)]. Cambridge University Press, Cambridge, UK, and New York, 976 pp. <http://www.ipcc.ch>

Kalnay, E., M. Kanamitsu, R. Kistler, W. Collins, D. Deaven, L. Gandin, M. Iredell, S. Saha, G. White, J. Woollen, Y. Zhu, A. Leetmaa, B. Reynolds, M. Chelliah, W. Ebisuzaki, W. Higgins, J. Janowiak, K.C. Mo, C. Ropelewski, J. Wang, R. Jenne, and D. Joseph, 1996: The NCEP/NCAR 40-Year Reanalysis Project. *Bulletin of the American Meteorological Society*, **77(3)**, 437-471.

Mesinger, F., G. DiMego, E. Kalnay, K. Mitchell, P.C. Shafran, W. Ebisuzaki, D. Jović, J. Woollen, E. Rogers, E.H. Berbery, M.B. Ek, Y. Fan, R. Grumbine, W. Higgins, H. Li, Y. Lin, G. Manikin, D. Parrish, and W. Shi, 2006: North American regional reanalysis. *Bulletin of the American Meteorological Society*, **87(3)**, 343-360.

NCDC (National Climatic Data Center), 2007: *Billion Dollar U.S. Weather Disasters, 1980-2006*. National Climatic Data Center, Asheville, NC. <http://www.ncdc.noaa.gov/oa/reports/billionz.html>

Pinto, J.G., E.L. Frohlich, G.C. Leckebusch and U. Ulbrich, 2007: Changing European storm loss potential under modified climate conditions according to ensemble simulations of the ECHAM5/MPI-OM1 GCM. *Natural Hazards and Earth System Sciences*, **7(1)**, 165-175.

Pryor, S.C., R.J. Barthelmie, and J.T. Schoof, 2006: Interannual variability of wind indices across Europe. *Wind Energy*, **9**, 27-38.

Pulwarty, R., 2003: Climate and water in the west: science, information and decision-making. *Water Resources Update*, **124**, 4-12.

Santer, B.D., T.M.L. Wigley, C. Mears, F.J. Wentz, S.A. Klein, D.J. Seidel, K.E. Taylor, P.W. Thorne, M.F. Wehner, P.J. Gleckler, J.S. Boyle, W.D. Collins, K.W. Dixon, C. Doutriaux, M. Free, Q. Fu, J.E. Hansen, G.S. Jones, R. Ruedy, T.R. Karl, J.R. Lanzante, G.A. Meehl, V. Ramaswamy, G. Russell, and G.A. Schmidt, 2005: Amplification of surface temperature trends and variability in the tropical atmosphere. *Science*, **309(5740)**, 1551-1556.

Schwartz, M.N. and R.L. George, 1998: *On the Use of Reanalysis Data for Wind Resource Assessment*. NREL/CP-500-25610. National Renewable Energy Laboratory, Golden, CO, 5 pp. <http://www.nrel.gov/docs/fy99osti/25610.pdf>

Stott, P.A., D.A. Stone, and M.R. Allen, 2004: Human contribution to the European heatwave of 2003. *Nature*, **432(7017)**, 610-614.

Webster's II Dictionary, 1988: *New Riverside University Dictionary*. Houghton Mifflin Co., Boston MA, 1535 pp.

CHAPTER 2 REFERENCES

AchutaRao, K.M., M. Ishii, B.D. Santer, P.J. Gleckler, K.E. Taylor, T.P. Barnett, D.W. Pierce, R.J. Stouffer, and T.M.L. Wigley, 2007: Simulated and observed variability in ocean temperature and heat content. *Proceedings of the National Academy of Sciences*, **104(26)**, 10768-10773.

Adler, R.F., G.J. Huffman, A. Chang, R. Ferraro, P. Xie, J. Janowiak, B. Rudolf, U. Schneider, S. Curtis, D. Bolvin, A. Gruber, J. Susskind, P. Arkin, and E. Nelkin, 2003: The Version-2 Global Precipitation Climatology Project (GPCP) monthly precipitation analysis (1979–present). *Journal of Hydrometeorology*, **4(6)**, 1147-1167.

Alpert, P., Y.J. Kaufman, Y. Shay-El, D. Tanre, A. da Silva, S. Schubert, and Y.H. Joseph, 1998: Quantification of dust-forced heating of the lower troposphere. *Nature*, **395(6700)**, 367-370.

Andersen, U.J., E. Kaas, and P. Alpert, 2001: Using analysis increments to estimate atmospheric heating rates following volcanic eruptions. *Geophysical Research Letters*, **28(6)**, 991-994.

Annan, J.D., J.C. Hargreaves, N.R. Edwards, and R. Marsh, 2005: Parameter estimation in an intermediate complexity earth system model using an ensemble Kalman filter. *Ocean Modelling*, **8(1-2)**, 135-154.

Arkin, P.A., 1982: The relationship between interannual variability in the 200 mb tropical wind field and the Southern Oscillation. *Monthly Weather Review*, **110(10)**, 1393-1404.

Baldwin, M.P. and T.J. Dunkerton, 1999: Downward propagation of the Arctic oscillation from the stratosphere to the troposphere. *Journal of Geophysical Research*, **104(D24)**, 30937-30946.

Baldwin, M.P. and T.J. Dunkerton, 2001: Stratospheric harbingers of anomalous weather regimes. *Science*, **294(5542)**, 581-584.

Barlow, M., S. Nigam, and E.H. Berbery, 2001: ENSO, Pacific decadal variability, and U.S. summertime precipitation, drought, and stream flow. *Journal of Climate*, **14(9)**, 2105-2128.

Basist, A.N. and M. Chelliah, 1997: Comparison of tropospheric temperatures derived from the NCEP/NCAR reanalysis, NCEP operational analysis, and the microwave sounding unit. *Bulletin of the American Meteorological Society*, **78(7)**, 1431-1447.

Beljaars, A., U. Andrae, P. Kallberg, A. Simmons, S. Uppala, and P. Viterbo, 2006: The hydrological cycle in atmospheric reanalysis. In: *Encyclopedia of Hydrological Sciences.* Part 15. Global Hydrology. [Online], John Wiley & Sons. doi:10.1002/0470848944.hsa189. Article online posting date: April 15, 2006.

Bengtsson, L. and J. Shukla, 1988: Integration of space and *in situ* observation to study climate change. *Bulletin of the American Meteorological Society*, **69(10)**, 1130-1143.

Bengtsson, L., S. Hagemann, and K.I. Hodges, 2004a: Can climate trends be calculated from reanalysis data? *Journal of Geophysical Research*, **109**, D11111, doi:10.1029/2004JD004536.

Bengtsson, L., K.I. Hodges, and S. Hagemann, 2004b: Sensitivity of large-scale atmospheric analyses to humidity observations and its impact on the global water cycle and tropical and extratropical weather systems in ERA40. *Tellus A*, **56(3)**, 202-217.

Bengtsson, L., K.I. Hodges, and S. Hagemann, 2004c: Sensitivity of the ERA40 reanalysis to the observing system: determination of the global atmospheric circulation from reduced observations. *Tellus A*, **56(3)**, 456-471.

Betts, A.K., P. Viterbo, and E. Wood, 1998a: Surface energy and water balance for the Arkansas–Red River basin from the EC-MWF reanalysis. *Journal of Climate*, **11(11)**, 2881-2897.

Betts, A.K., P. Viterbo, and A.C.M. Beljaars, 1998b: Comparison of the land-surface interaction in the ECMWF reanalysis model with the 1987 FIFE data. *Monthly Weather Review*, **126(1)**, 186-198.

Betts, A.K., J.H. Ball, P. Viterbo, A. Dai, and J. Marengo, 2005: Hydrometeorology of the Amazon in ERA-40. *Journal of Hydrometeorology*, **6(5)**, 764-774.

Blackmon, M.L., J.M. Wallace, N.-C. Lau, and S.L. Mullen 1977: An observational study of the Northern Hemisphere wintertime circulation. *Journal of the Atmospheric Sciences*, **34(7)**, 1040-1053.

Bosilovich, M.G., S.D. Schubert, M. Rienecker, R. Todling, M. Suarez, J. Bacmeister, R. Gelaro, G.-K. Kim, I. Stajner, and J. Chen, 2006: NASA's Modern Era Retrospective-analysis for Research and Applications (MERRA). *U.S. CLIVAR Variations*, **4(2)**, 5-8.

Boyer, T.P., J.I. Antonov, H. Garcia, D.R. Johnson, R.A. Locarnini, A.V. Mishonov, M.T. Pitcher, O.K. Baranova, and I. Smolyar, 2006: *World Ocean Database 2005* [Levitus, S. (ed.)]. NOAA Atlas NESDIS 60. National Environmental Satellite, Data, and Information Service, Silver Spring, MD, 182 pp

Bradley, R.S., M.K. Hughes, and H.F. Diaz, 2003: Climate in Medieval time. *Science*, **302(5644)**, 404-405.

Bromwich, D.H. and S.-H. Wang, 2005: Evaluation of the NCEP-NCAR and ECMWF 15- and 40-yr reanalyses using rawinsonde data from two independent Arctic field experiments. *Monthly Weather Review*, **133(12)**, 3562-3578.

Bromwich, D.H., R.L. Fogt, K.I. Hodges, and J.E. Walsh, 2007: A tropospheric assessment of the ERA-40, NCEP, and JRA-25 global reanalyses in the polar regions. *Journal of Geophysical Research*, **112**, D10111, doi:10.1029/2006JD007859.

Brönnimann, S., G.P. Compo, P.D. Sardeshmukh, R. Jenne, and S. Sterin, 2005: New approaches for extending the twentieth century climate record. *EOS, Transactions of the American Geophysical Union*, **86(1)**, 2, 6.

Carton, J.A. and B.S. Giese, 2008: A reanalysis of ocean climate using Simple Ocean Data Assimilation (SODA). *Monthly Weather Review*, **136(8)**, 2999-3017.

Carton, J.A., G. Chepurin, X. Cao, and B.S. Giese, 2000: A simple ocean data assimilation analysis of the global upper ocean 1950–95. Part I: ethodology. *Journal of Physical Oceanography*, **30(2)**, 294-309.

Cash, B.A. and S. Lee, 2001: Observed nonmodal growth of the Pacific-North American teleconnection pattern. *Journal of Climate,* **14(6)**, 1017-1028.

CCSP (Climate Change Science Program), 2006: *Temperature Trends in the Lower Atmosphere: Steps for Understanding and Reconciling Differences.* [Karl, T.R., S.J. Hassol, C.D. Miller, and W.L. Murray (eds.)]. Synthesis and Assessment Product 1.1. U.S. Climate Change Science Program, Washington, DC, 164 pp.

Charney, J.G., 1951: Dynamic forecasting by numerical process. In: *Compendium of Meteorology* [Malone, T.F. (ed.)]. American Meteorological Society, Boston, pp. 470-482.

Chelliah, M. and G.D. Bell, 2004: Tropical multidecadal and interannual climate variability in the NCEP–NCAR reanalysis. *Journal of Climate,* **17(9)**, 1777-1803.

Chelliah, M. and C. Ropelewski, 2000: Reanalyses-based tropospheric temperature estimates: uncertainties in the context of global climate change detection. *Journal of Climate,* **13(17)**, 3187-3205

Chen, J. and M.G. Bosilovich, 2007: Hydrological variability and trends in global reanalyses. In: *19th Conference on Climate Variability and Change,* San Antonio, Texas, January 2007. American Meteorological Society, (Boston), paper JP4.3. <http://ams.confex.com/ams/pdfpapers/119754.pdf>

Chepurin, G.A., J.A. Carton, and D. Dee, 2005: Forecast model bias correction in ocean data assimilation. *Monthly Weather Review,* **133(5)**, 1328-1342.

Collins, W.D., V. Ramaswamy, M.D. Schwarzkopf, Y. Sun, R.W. Portmann, Q. Fu, S.E.B. Casanova, J.-L. Dufresne, D.W. Fillmore, P.M.D. Forster, V.Y. Galin, L.K. Gohar, W.J. Ingram, D.P. Kratz, M.-P. Lefebvre, J. Li, P. Marquet, V. Oinas, Y. Tsushima, T. Uchiyama, and W.Y. Zhong, 2006: Radiative forcing by well-mixed greenhouse gases: estimates from climate models in the Intergovernmental Panel on Climate Change (IPCC) Fourth Assessment Report (AR4). *Journal of Geophysical Research,* **111**, D14317, doi:10.1029/2005JD006713.

Compo, G.P., P.D. Sardeshmukh, and C. Penland, 2001: Changes of subseasonal variability associated with El Niño. *Journal of Climate,* **14(16)**, 3356-3374.

Compo, G.P., J.S. Whitaker, and P.D. Sardeshmukh, 2006: Feasibility of a 100 year reanalysis using only surface pressure data. *Bulletin of the American Meteorological Society,* **87(2)**, 175-190.

Cullather, R.I., D.H. Bromwich, and M.C. Serreze, 2000: The atmospheric hydrologic cycle over the Arctic basin from reanalyses. Part I: Comparison with observations and previous studies. *Journal of Climate,* **13(5)**, 923-937.

Danforth, C.M., E. Kalnay, and T. Miyoshi, 2007: Estimating and correcting global weather model errors. *Monthly Weather Review,* **135(2)**, 281-299.

Dee, D.P., 2005: Bias and data assimilation. *Quarterly Journal of the Royal Meteorological Society,* **131(613)**, 3323-3343.

Dee, D.P. and A.M. da Silva, 1998: Data assimilation in the presence of forecast bias. *Quarterly Journal of the Royal Meteorological Society,* **124**, 269-295.

Dee, D.P. and R. Todling, 2000: Data assimilation in the presence of forecast bias: The GEOS moisture analysis. *Monthly Weather Review,* **128(9)**, 3268-3282.

Delworth, T.L., A.J. Broccoli, A. Rosati, R.J. Stouffer, V. Balaji, J.A. Beesley, W.F. Cooke, K.W. Dixon, J. Dunne, K.A. Dunne, J.W. Durachta, K.L. Findell, P. Ginoux, A. Gnanadesikan, C.T. Gordon, S.M. Griffies, R. Gudgel, M.J. Harrison, I.M. Held, R.S. Hemler, L.W. Horowitz, S.A. Klein, T.R. Knutson, P.J. Kushner, A.R. Langenhorst, H.C. Lee, S.J. Lin, J. Lu, S.L. Malyshev, P.C.D. Milly, V. Ramaswamy, J. Russell, M.D. Schwarzkopf, E. Shevliakova, J.J. Sirutis, M.J. Spelman, W.F. Stern, M. Winton, A.T. Wittenberg, B. Wyman, F. Zeng, and R. Zhang, 2006: GFDL's CM2 global coupled climate models. Part I: formulation and simulation characteristics. *Journal of Climate,* **19(5)**, 643-674.

DeWeaver, E. and S. Nigam, 2002: Linearity in ENSO's atmospheric response. *Journal of Climate,* **15(17)**, 2446-2461.

Dirmeyer, P.A., X. Gao, M. Zhao, Z. Guo, T. Oki, and N. Hanasaki, 2006: GSWP-2, Multimodel analysis and implications for our perception of the land surface. *Bulletin of the American Meteorological Society,* **87(10)**, 1381-1397.

Doherty, R.M., M. Hulme, and C.G. Jones, 1999: A gridded reconstruction of land and ocean precipitation for the extended tropics from 1974 to 1994. *International Journal of Climatology,* **19(2)**, 119-142.

Entekhabi, D., E.G. Njoku, P. Houser, M. Spencer, T. Doiron, Y. Kim, J. Smith, R. Girard, S. Belair, W. Crow, T.J. Jackson, Y.H. Kerr, J.S. Kimball, R. Koster, K.C. McDonald, P.E. O'Neill, T. Pultz, S.W. Running, J. Shi, E. Wood, and J. van Zyl, 2004: The Hydrosphere State (Hydros) satellite mission: an earth system pathfinder for global mapping of soil moisture and land freeze/thaw. *IEEE Transactions on Geoscience and Remote Sensing,* **42(10)**, 2184-2195.

Fan, Y. and H. van den Dool, 2008: A global monthly land surface air temperature analysis for 1948–present. *Journal of Geophysical Research,* **113**, D01103, doi:10.1029/2007JD008470.

Farrell, B.F., 1989: Optimal excitation of baroclinic waves. *Journal of the Atmospheric Sciences,* **46(9)**, 1193-1206.

Feldstein, S.B., 2002: Fundamental mechanisms of the growth and decay of the PNA teleconnection pattern. *Quarterly Journal of the Royal Meteorological Society,* **128(581)**, 775-796.

Feldstein, S.B., 2003: The dynamics of NAO teleconnection pattern growth and decay. *Quarterly Journal of the Royal Meteorological Society,* **129(589)**, 901-924.

Folland, C.K., T.N. Palmer, and D.E. Parker, 1986: Sahel rainfall and worldwide sea temperatures, 1901-85. *Nature,* **320(6063)**, 602-607.

Foster, J.L., C. Sun, J.P. Walker, R. Kelly, A. Chang, J. Dong, and H. Powell, 2005: Quantifying the uncertainty in passive microwave snow water equivalent observations. *Remote Sensing of Environment,* **94(2)**, 187-203.

Gaffen, D.J., 1994: Temporal inhomogeneities in radiosonde temperature records. *Journal of Geophysical Research,* **99(D2)**, 3667-3676.

Gates, W.L., 1992: AMIP: The Atmospheric Model Intercomparison Project. *Bulletin of the American Meteorological Society,* **73(12)**, 1962-1970.

Gates, W.L., J.S. Boyle, C. Covey, C.G. Dease, C.M. Doutriaux, R.S. Drach, M. Fiorino, P.J. Gleckler, J.J. Hnilo, S.M. Marlais, T.J. Phillips, G.L. Potter, B.D. Santer, K.R. Sperber, K.E. Taylor, and D.N. Williams, 1999: An overview of the results of the Atmospheric Model Intercomparison Project (AMIP I). *Bulletin of the American Meteorological Society,* **80(1)**, 29-55.

GEOSS (Global Earth Observation System of Systems), 2005: *The Global Earth Observation System of Systems (GEOSS) 10-Year Implementation Plan.* [GEO Secretariat, Geneva, Switzerland], 11 pp. <http://www.earthobservations.org/documents/10-Year%20Implementation%20Plan.pdf>

Gibson, J.K., P. Kållberg, S. Uppala, A. Hernandez, A. Nomura, and E. Serrano, 1997: *ERA description.* ERA-15 Project report series 1. European Centre for Medium-Range Weather Forecasts, Reading (UK), 72 pp.

Grassl, H., 2000: Status and improvements of coupled general circulation models. *Science,* **288(5473)**, 1991-1997.

Groisman, P.Ya., R.W. Knight, T.R. Karl, D.R. Easterling, B. Sun, and J.H. Lawrimore, 2004: Contemporary changes of the hydrological cycle over the contiguous United States: trends derived from *in situ* observations. *Journal of Hydrometeorology,* **5(1)**, 64-85.

Gulev, S.K., O. Zolina, and S. Grigoriev. 2001. Extratropical cyclone variability in the Northern Hemisphere winter from the NCEP/NCAR reanalysis data. *Climate Dynamics,* **17(10)**, 795-809.

Gutzler, D.S., R.D. Rosen, D.A. Salstein, and J.P. Peixoto, 1988: Patterns of interannual variability in the Northern Hemisphere wintertime 850 mb temperature field. *Journal of Climate,* **1(10)**, 949-964.

Haimberger, L., 2007: Homogenization of radiosonde temperature time series using innovation statistics, *Journal of Climate,* **20(7)**, 1377-1403.

Hamlet, A.F. and D.P. Lettenmaier, 2005: Production of temporally consistent gridded precipitation and temperature fields for the continental United States. *Journal of Hydrometeorology,* **6(3)**, 330-336.

Hanawa, K., P. Rual, R. Bailey, A. Sy, and M. Szabados, 1995: A new depth-time equation for Sippican or TSK T-7, T-6 and T-4 expendable bathythermographs (XBT). *Deep-Sea Research Part I: Oceanographic Research Papers,* **42(8)**, 1423-1452.

Hansen, J., D. Johnson, A. Lacis, S. Lebedeff, P. Lee, D. Rind, and G. Russell, 1981: Climate impact of increasing atmospheric carbon dioxide. *Science,* **213(4511)**, 957-966.

Hansen, J.E., R. Ruedy, M. Sato, M. Imhoff, W. Lawrence, D. Easterling, T. Peterson, and T. Karl, 2001: A closer look at United States and global surface temperature change. *Journal of Geophysical Research,* **106(D20)**, 23947-23963.

Hays, J.D., J. Imbrie, and N.J. Shackleton, 1976: Variations in the Earth's orbit: pacemaker of the ice ages. *Science,* **194(4270)**, 1121-1132.

Higgins, R.W., K.C. Mo, and S.D. Schubert, 1996: The moisture budget of the central United States in spring as evaluated in the NCEP/NCAR and the NASA/DAO reanalyses. *Monthly Weather Review,* **124(5)**, 939-963.

Hoaglin, D.C., F. Mosteller, and J.W. Tukey, 1983: *Understanding Robust and Exploratory Data Analysis.* Wiley and Sons, New York, 447 pp.

Hodges, K.I., B.J. Hoskins, J. Boyle, and C. Thorncroft, 2003: A comparison of recent reanalysis datasets using objective feature tracking: storm tracks and tropical easterly waves. *Monthly Weather Review,* **131(9)**, 2012-2036.

Hoerling, M.P. and A. Kumar, 2002: Atmospheric response patterns associated with tropical forcing. *Journal of Climate,* **15(16)**, 2184-2203.

Hoerling, M.P. and A. Kumar, 2003: The perfect ocean for drought. *Science,* **299(5607)**, 691-699.

Huffman, G.J., R.F. Adler, P. Arkin, A. Chang, R. Ferraro, A. Gruber, J. Janowiak, A. McNab, B. Rudolf, and U. Schneider, 1997: The Global Precipitation Climatology Project (GPCP) combined precipitation dataset. *Bulletin of the American Meteorological Society,* **78(1)**, 5-20.

Hurrell, J.W., 1996: Influence of variations in extratropical wintertime teleconnections on Northern Hemisphere temperature. *Geophysical Research Letters,* **23(6),** 665-668.

Hurrell, J.W., Y. Kushnir, and M. Visbeck, 2001: The North Atlantic Oscillation. *Science,* **291(5504),** 603-605.

IPCC (Intergovernmental Panel on Climate Change) 2007: *Climate Change 2007: The Physical Science Basis.* Contribution of Working Group I to the Fourth Assessment Report (AR4) of the Intergovernmental Panel on Climate Change [Solomon, S., D. Qin, M. Manning, Z. Chen, M. Marquis, K.B. Averyt, M. Tignor, and H.L. Miller (eds.)]. Cambridge University Press, Cambridge, UK, and New York, 996 pp. <http://www.ipcc.ch>

Jeuken, A.B.M., P.C. Siegmund, L.C. Heijboer, J. Feichter, and L. Bengtsson, 1996: On the potential of assimilating meteorological analyses in a global climate model for the purpose of model validation. *Journal of Geophysical Research,* **101(D12),** 16939-16950.

Jones, P.D., 1994a: Hemispheric surface air temperature variations: A reanalysis and an update to 1993. *Journal of Climate,* **7(11),** 1794-1802.

Jones, P.D., 1994b: Recent warming in global temperature series. *Geophysical Research Letters,* **21(12),** 1149-1152.

Jones, P.D. and A. Moberg, 2003: Hemispheric and large-scale surface air temperature variations: An extensive revision and an update to 2001. *Journal of Climate,* **16(2),** 206-223.

Jones, P.D., M. New, D.E. Parker, S. Martin, and I.G. Rigor, 1999: Surface air temperature and its changes over the past 150 years. *Reviews of Geophysics,* **37(2),** 173-199.

Kaas, E., A. Guldberg, W. May, and M. Déqué, 1999: Using tendency errors to tune the parameterization of unresolved dynamical scale interactions in atmospheric general circulation models. *Tellus A,* **51(5),** 612-629.

Kalnay, E., 2003: *Atmospheric Modeling, Data Assimilation and Predictability.* New York, Cambridge University Press, 341 pp.

Kalnay, E. and M. Cai, 2003: Impact of urbanization and land-use change on climate. *Nature,* **423(6939),** 528-531.

Kalnay, E., M. Kanamitsu, R. Kistler, W. Collins, D. Deaven, L. Gandin, M. Iredell, S. Saha, G. White, J. Woollen, Y. Zhu, A. Leetmaa, B. Reynolds, M. Chelliah, W. Ebisuzaki, W. Higgins, J. Janowiak, K.C. Mo, C. Ropelewski, J. Wang, R. Jenne, and D. Joseph, 1996: The NCEP/NCAR 40-Year Reanalysis Project. *Bulletin of the American Meteorological Society,* **77(3),** 437-471.

Kalnay, E., M. Cai, H. Li, and J. Tobin, 2006: Estimation of the impact of land-surface forcings on temperature trends in eastern United States. *Journal of Geophysical Research,* **111,** D06106, doi:10.1029/2005JD006555.

Kanamitsu, M. and S.-O. Hwang, 2006: Role of sea surface temperature in reanalysis. *Monthly Weather Review,* **134(2),** 532-552.

Kanamitsu, M. and S. Saha, 1996: Systematic tendency error in budget calculations. *Monthly Weather Review,* **124(6),** 1145-1160.

Kanamitsu, M., W. Ebisuzaki, J. Woollen, S.-K. Yang, J.J. Hnilo, M. Fiorino, and G.L. Potter, 2002: NCEP–DOE AMIP-II Reanalysis (R-2). *Bulletin of the American Meteorological Society,* **83(11),** 1631-1643.

Karl, T.R. and R.W. Knight, 1998: Secular trends of precipitation amount, frequency, and intensity in the United States. *Bulletin of the American Meteorological Society,* **79(2),** 231-241.

Karl, T.R., G. Kukla, V. Razuvayev, M. Changery, R.G. Quayle, R.R. Heim, D.R. Easterling, and C.B. Fu, 1991: Global warming: evidence for asymmetric diurnal temperature change. *Geophysical Research Letters,* **18(12),** 2253-2256.

Kinter, J.L., M.J. Fennessy, V. Krishnamurthy, and L. Marx, 2004: An evaluation of the apparent interdecadal shift in the tropical divergent circulation in the NCEP-NCAR reanalysis. *Journal of Climate,* **17(2),** 349-361.

Kistler, R., E. Kalnay, W. Collins, S. Saha, G. White, J. Woollen, M. Chelliah, W. Ebisuzaki, M. Kanamitsu, V. Kousky, H. van den Dool, R. Jenne, and M. Fiorino, 2001: The NCEP–NCAR 50–year reanalysis: monthly means CD–ROM and documentation. *Bulletin of the American Meteorological Society,* **82(2),** 247-267.

Klinker, E. and P.D. Sardeshmukh, 1992: The diagnosis of mechanical dissipation in the atmosphere from large-scale balance requirements. *Journal of the Atmospheric Sciences,* **49(7),** 608-626.

Lau, N.-C., H. Tennekes, and J.M. Wallace, 1978: Maintenance of the momentum flux by transient eddies in the upper troposphere. *Journal of the Atmospheric Sciences,* **35(1),** 139-147.

Lee, M.-I., S.D. Schubert, M.J. Suarez, I.M. Held, N.-C. Lau, J.J. Ploshay, A. Kumar, H.-K. Kim, and J.-K.E. Schemm, 2007: An analysis of the warm-season diurnal cycle over the continental United States and northern Mexico in general circulation models. *Journal of Hydrometeorology,* **8(3),** 344-366.

Lin, J., B. Mapes, M. Zhang, and M. Newman, 2004: Stratiform precipitation, vertical heating profiles, and the Madden–Julian Oscillation. *Journal of the Atmospheric Sciences,* **61(3),** 296-309.

Liu, A.Z., M. Ting, and H. Wang, 1998: Maintenance of circulation anomalies during the 1988 drought and 1993 floods over the United States. *Journal of the Atmospheric Sciences,* **55(17)**, 2810-2832.

Luo, Y., E.H. Berbery, K.E. Mitchell, and A.K. Betts, 2007: Relationships between land surface and near-surface atmospheric variables in the NCEP North American regional reanalysis. *Journal of Hydrometeorology,* **8(6)**, 1184-1203.

Madden, R.A. and P.R. Julian, 1994: Observations of the 40-50 day tropical oscillation: a review. *Monthly Weather Review,* **122(5)**, 814-837.

Massacand, A.C. and H.C. Davies, 2001: Interannual variability of the extratropical northern hemisphere and the potential vorticity wave guide. *Atmospheric Science Letters*, **2(1-4)**, 61-71.

Mears, C.A., M.C. Schabel, and F.J. Wentz, 2003: A reanalysis of the MSU channel 2 tropospheric temperature record. *Journal of Climate,* **16(22)**, 3650-3664.

Mesinger, F., G. DiMego, E. Kalnay, K. Mitchell, P.C. Shafran, W. Ebisuzaki, D. Jović, J. Woollen, E. Rogers, E.H. Berbery, M.B. Ek, Y. Fan, R. Grumbine, W. Higgins, H. Li, Y. Lin, G. Manikin, D. Parrish, and W. Shi, 2006: North American regional reanalysis. *Bulletin of the American Meteorological Society,* **87(3)**, 343-360.

Mitas, C.M. and A. Clement, 2006: Recent behavior of the Hadley cell and tropical thermodynamics in climate models and reanalyses. *Geophysical Research Letters*, **33**, L01810, doi:10.1029/2005GL024406.

Mitchell, K.E., D. Lohmann, P.R. Houser, E.F. Wood, J.C. Schaake, A. Robock, B.A. Cosgrove, J. Sheffield, Q. Duan, L. Luo, R.W. Higgins, R.T. Pinker, J.D. Tarpley, D.P. Lettenmaier, C.H. Marshall, J.K. Entin, M. Pan, W. Shi, V. Koren, J. Meng, B.H. Ramsay, and A.A. Bailey, 2004: The multi-institution North American Land Data Assimilation System (NLDAS): Utilizing multiple GCIP products and partners in a continental distributed hydrological modeling system. *Journal of Geophysical Research,* **109**, D07S90, doi:10.1029/2003JD003823.

Mo, K.C., 2000: Relationships between low-frequency variability in the Southern Hemisphere and sea surface temperature anomalies. *Journal of Climate,* **13(20)**, 3599-3610.

Mo, K.C., J.N. Paegle, and R.W. Higgins, 1997: Atmospheric processes associated with summer floods and droughts in the central United States. *Journal of Climate,* **10(12)**, 3028-3046.

Newman, M., P.D. Sardeshmukh, and J.W. Bergman, 2000: An assessment of the NCEP, NASA, and ECMWF reanalyses over the tropical west Pacific warm pool. *Bulletin of the American Meteorological Society*, **81(1)**, 41-48.

Nigam, S. and A. Ruiz-Barradas, 2006: Seasonal hydroclimate variability over North America in global and regional reanalyses and AMIP simulations: varied representation. *Journal of Climate,* **19(5)**, 815-837.

Nigam S., C. Chung, and E. DeWeaver, 2000: ENSO diabatic heating in ECMWF and NCEP–NCAR reanalyses, and NCAR CCM3 simulation. *Journal of Climate,* **13(17)**, 3152-3171.

Onogi, K., H. Koide, M. Sakamoto, S. Kobayashi, J. Tsutsui, H. Hatsushika, T. Matsumoto, N. Yamazaki, H. Kamahori, K. Takahashi, K. Kato, R. Oyama, T. Ose, S. Kadokura, K. Wada, 2005: JRA-25: Japanese 25-year re-analysis–progress and status. *Quarterly Journal of the Royal Meteorological Society*, **131(613)**, 3259-3268.

Ott, E., B.R Hunt, I. Szunyogh, A.V. Zimin, E.J. Kostelich, M. Corazza, E. Kalnay, D.J. Patil, and J.A. Yorke, 2004: A local ensemble Kalman filter for atmospheric data assimilation. *Tellus A*, **56(5)**, 415-428.

Palmer, T.N., 1988: Medium and extended range predictability, and the stability of the PNA mode. *Quarterly Journal of the Royal Meteorological Society,* **114(481)**, 691-713.

Pavelsky, T.M. and L.C. Smith, 2006: Intercomparison of four global precipitation datasets and their correlation with increased Eurasian river discharge to the Arctic Ocean. *Journal of Geophysical Research,* **111**, D21112, doi:10.1029/2006JD007230.

Pawson, S. and M. Fiorino, 1998a: A comparison of reanalyses in the tropical stratosphere. Part 1: thermal structure and the annual cycle. *Climate Dynamics,* **14(9)**, 631-644.

Pawson, S. and M. Fiorino, 1998b: A comparison of reanalyses in the tropical stratosphere. Part 2: the quasi-biennial oscillation. *Climate Dynamics,* **14(9)**, 645-658.

Pawson, S. and M. Fiorino, 1999: A comparison of reanalyses in the tropical stratosphere. Part 3: inclusion of the pre-satellite data era. *Climate Dynamics,* **15(4)**, 241-250.

Pawson, S., K. Kodera, K. Hamilton, T.G. Shepherd, S.R. Beagley, B.A. Boville, J.D. Farrara, T.D.A. Fairlie, A. Kitoh, W.A. Lahoz, U. Langematz, E. Manzini, D.H. Rind, A.A. Scaife, K. Shibata, P. Simon, R. Swinbank, L. Takacs, R.J. Wilson, J.A. Al-Saadi, M. Amodei, M. Chiba, L. Coy, J. de Grandpré, R.S. Eckman, M. Fiorino, W.L. Grose, H. Koide, J.N. Koshyk, D. Li, J. Lerner, J.D. Mahlman, N.A. McFarlane, C.R. Mechoso, A. Molod, A. O'Neill, R.B. Pierce, W.J. Randel, R.B. Rood, and F. Wu, 2000: The GCM–Reality Intercomparison Project for SPARC (GRIPS): Scientific issues and initial results. *Bulletin of the American Meteorological Society,* **81(4)**, 781-796.

Philander, S.G., 1990: *El Niño, La Niña, and the Southern Oscillation*. Academic Press, San Diego, CA, 293 pp.

Pielke, R.A., R.L. Walko, L.T. Steyaert, P.L. Vidale, G.E. Liston, W.A. Lyons, and T.N. Chase, 1999: The influence of anthropogenic landscape changes on weather in south Florida. *Monthly Weather Review*, **127(7)**, 1663-1673.

Pohlmann, H. and R.J. Greatbatch, 2006: Discontinuities in the late 1960's in different atmospheric data products. *Geophysical Research Letters*, **33**, L22803, doi:10.1029/2006GL027644.

Raible, C.C., 2007: On the relation between extremes of midlatitude cyclones and the atmospheric circulation using ERA40. *Geophysical Research Letters*, **34**, L07703, doi:10.1029/2006GL029084.

Randall, D. (ed.), 2000: *General Circulation Model Development: Past, Present and Future*. International geophysics series v. 70. Academic Press, San Diego, 807 pp.

Reichle, R.H. and R.D. Koster, 2005: Global assimilation of satellite surface soil moisture retrievals into the NASA Catchment land surface model. *Geophysical Research Letters*, **32**, L02404, doi:10.1029/2004GL021700.

Reichle, R.H., R.D. Koster, P. Liu, S.P.P. Mahanama, E.G. Njoku, and M. Owe, 2007: Comparison and assimilation of global soil moisture retrievals from the Advanced Microwave Scanning Radiometer for the Earth Observing System (AMSR-E) and the Scanning Multichannel Microwave Radiometer (SMMR). *Journal of Geophysical Research*, **112**, D09108, doi:10.1029/2006JD008033.

Richardson, L.F., 1922: *Weather Prediction by Numerical Process*. Cambridge University Press, reprinted by Dover Publications, New York, 1965, p. 46.

Robock, A., K.Y. Vinnikov, G. Srinivasan, J.K. Entin, S.E. Hollinger, N.A. Speranskaya, S. Liu, and A. Namkhai, 2000: The global soil moisture data bank. *Bulletin of the American Meteorological Society*, **81(6)**, 1281-1299.

Rodell, M., P.R. Houser, U. Jambor, J. Gottschalck, K. Mitchell, C.-J. Meng, K. Arsenault, B. Cosgrove, J. Radakovich, M. Bosilovich, J.K. Entin, J.P. Walker, D. Lohmann, and D. Toll, 2004: The global land data assimilation system. *Bulletin of the American Meteorological Society*, **85(3)**, 381-394.

Rodell, M., J.L. Chen, H. Kato, J.S. Famiglietti, J. Nigro, and C.R. Wilson, 2007: Estimating groundwater storage changes in the Mississippi River basin (USA) using GRACE. *Hydrogeology Journal*, **15(1)**, 159-166.

Rodwell, M.J. and T.N. Palmer, 2007: Using numerical weather prediction to assess climate models. *Quarterly Journal of the Royal Meteorological Society*, **133(622)**, 129-146.

Roemmich, D. and W.B. Owens, 2000: The Argo Project: global ocean observations for understanding and prediction of climate variability. *Oceanography*, **13(2)**, 45-50.

Ruiz-Barradas, A. and S. Nigam, 2005: Warm-season rainfall variability over the U.S. Great Plains in observations, NCEP and ERA-40 reanalyses, and NCAR and NASA atmospheric model simulations. *Journal of Climate*, **18(11)**, 1808-1829.

Rusticucci, M. and V.E. Kousky, 2002: A comparative study of maximum and minimum temperatures over Argentina: NCEP/NCAR reanalysis versus station data. *Journal of Climate*, **15(15)**, 2089-2101.

Saha, S., S. Nadiga, C. Thiaw, J. Wang, W. Wang, Q. Zhang, H.M. Van den Dool, H.-L. Pan, S. Moorthi, D. Behringer, D. Stokes, M. Pena, S. Lord, G. White, W. Ebisuzaki, P. Peng, and P. Xie, 2006: The NCEP climate forecast system. *Journal of Climate*, **19(15)**, 3483-3517.

Santer, B.D., R. Sausen, T.M.L. Wigley, J.S. Boyle, K. AchutaRao, C. Doutriaux, J.E. Hansen, G.A. Meehl, E. Roeckner, R. Ruedy, G. Schmidt, and K.E. Taylor, 2003: Behavior of tropopause height and atmospheric temperature in models, reanalyses, and observations: decadal changes. *Journal of Geophysical Research*, **108(D1)**, 4002, doi:10.1029/2002JD002258.

Santer, B.D., T.M.L. Wigley, A.J. Simmons, P.W. Kållberg, G.A. Kelly, S.M. Uppala, C. Ammann, J.S. Boyle, W. Brüggemann, C. Doutriaux, M. Fiorino, C. Mears, G.A. Meehl, R. Sausen, K.E. Taylor, W.M. Washington, M.F. Wehner, and F.J. Wentz, 2004: Identification of anthropogenic climate change using a second-generation reanalysis. *Journal of Geophysical Research*, **109(D2)**, 1104, doi:10.1029/2004JD005075.

Sardeshmukh, P.D., 1993: The baroclinic χ [chi] problem and its application to the diagnosis of atmospheric heating rates. *Journal of the Atmospheric Sciences*, **50(8)**, 1099-1112.

Sardeshmukh, P.D., G.P. Compo, and C. Penland, 2000: Changes of probability associated with El Niño. *Journal of Climate*, **13(24)**, 4268-4286.

Schubert, S. and Y. Chang, 1996: An objective method for inferring sources of model error. *Monthly Weather Review*, **124(2)**, 325-340.

Schubert, S.D., R. Rood, and J. Pfaendtner, 1993: An assimilated dataset for earth science applications. *Bulletin of the American Meteorological Society*, **74(12)**, 2331-2342.

Schubert S.D., M.J. Suarez, P.J. Pegion, R.D. Koster, and J.T. Bacmeister. 2004: On the cause of the 1930s dust bowl. *Science*, **303(5665)**, 1855-1859.

Schubert, S., D. Dee, S. Uppala, J. Woollen, J. Bates, and S. Worley, 2006: *Report of the Workshop on "The Development of Improved Observational Datasets for Reanalysis: Lessons Learned and Future Directions"*. University of Maryland Conference Center, College Park, MD, 28-29 September 2005. [NASA Global Modeling and Assimilation Office, Greenbelt,

MD], 31 pp. <http://gmao.gsfc.nasa.gov/pubs/docs/Schubert273.pdf.>

Schubert, S.D., Y. Chang, M.J. Suarez, and P.J. Pegion, 2008: ENSO and wintertime extreme precipitation events over the contiguous United States. *Journal of Climate,* **21(1)**, 22-39.

Seager, R., Y. Kushnir, C. Herweijer, N. Naik, and J. Velez: 2005: Modeling of tropical forcing of persistent droughts and pluvials over western North America: 1856-2000. *Journal of Climate,* **18(19)**, 4065-4088.

Serreze, M.C., J.R. Key, J.E. Box, J.A. Maslanik, and K. Steffen, 1998: A new monthly climatology of global radiation for the Arctic and comparisons with NCEP–NCAR reanalysis and ISCCP-C2 fields. *Journal of Climate,* **11(2)**, 121-136.

Shinoda, T., H.H. Hendon, and J. Glick, 1999: Intraseasonal surface fluxes in the tropical western Pacific and Indian oceans from NCEP reanalyses. *Monthly Weather Review,* **127(5)**, 678-693.

Simmons, A.J., 2006: Observations, assimilation and the improvement of global weather prediction - some results from operational forecasting and ERA-40. In: *Predictability of Weather and Climate* [Palmer, T. and R. Hagedorn (eds.)], Cambridge University Press, Cambridge, UK, and New York, pp. 428-458.

Simmons, A.J., P.D. Jones, V. da Costa Bechtold, A.C.M. Beljaars, P.W. Kållberg, S. Saarinen, S.M. Uppala, P. Viterbo, and N. Wedi, 2004: Comparison of trends and low frequency variability in CRU, ERA-40, and NCEP/NCAR analyses of surface air temperature. *Journal of Geophysical Research,* **109**, D24115, doi:10.1029/2004JD005306.

Smith, R.D., M.E. Maltrud, F.O. Bryan, and M.W. Hecht, 2000: Numerical simulation of the North Atlantic Ocean at 1/10°. *Journal of Physical Oceanography,* **30(7)**, 1532-1561.

Stendel, M., J.R. Christy, and L. Bengtsson, 2000: Assessing levels of uncertainty in recent temperature time series. *Climate Dynamics,* **16(8)**, 587-601.

Straus, D.M. and J. Shukla, 2002: Does ENSO force the PNA? *Journal of Climate,* **15(17)**, 2340-2358.

Sun, C., J.P. Walker, and P.R. Houser, 2004: A methodology for snow data assimilation in a land surface model. *Journal of Geophysical Research,* **109**, D08108, doi:10.1029/2003JD003765.

Takahashi, K., N. Yamazaki, and H. Kamahori, 2006: Trends of heavy precipitation events in global observation and reanalysis datasets. *SOLA (Scientific Online Letters on the Atmosphere),* **2**, 96-99, doi:10.2151/sola.2006-025.

Thompson, D.W. and J.M. Wallace, 2000: Annular modes in the extratropical circulation. Part I: Month-to-month variability. *Journal of Climate,* **13(5)**, 1000-1016.

Tian, B., D.E. Waliser, E.J. Fetzer, B.H. Lambrigtsen, Y. Yung, and B. Wang, 2006: Vertical moist thermodynamic structure and spatial-temporal evolution of the MJO in AIRS observations. *Journal of the Atmospheric Sciences,* **63(10)**, 2462-2485.

Tippett, M.K., J.L. Anderson, C.H. Bishop, T.M. Hamill, and J.S. Whitaker, 2003: Ensemble square root filters. *Monthly Weather Review,* **131(7)**, 1485-1490.

Trenberth, K.E. and G.W. Branstator, 1992: Issues in establishing causes of the 1988 drought over North America. *Journal of Climate,* **5(2)**, 159-172.

Trenberth, K.E. and J.M. Caron, 2000: The Southern Oscillation revisited: sea level pressures, surface temperatures, and precipitation. *Journal of Climate,* **13(24)**, 4358-4365.

Trenberth, K.E. and J.M. Caron, 2001: Estimates of meridional atmosphere and ocean heat transports. *Journal of Climate,* **14(16)**, 3433-3443.

Trenberth, K.E. and C.J. Guillemot, 1996: Physical processes involved in the 1988 drought and 1993 floods in North America. *Journal of Climate,* **9(6)**, 1288-1298.

Trenberth K.E. and C.J. Guillemot, 1998: Evaluation of the atmospheric moisture and hydrological cycle in the NCEP/NCAR reanalyses. *Climate Dynamics,* **14(3)**, 213-231.

Trenberth, K.E. and J.G. Olson, 1988: An evaluation and intercomparison of global analyses from National Meteorological Center and European Centre for Medium Range Weather Forecasting. *Bulletin of the American Meteorological Society,* **69(9)**, 1047-1057.

Trenberth, K.E. and D.J. Shea, 2005: Relationships between precipitation and surface temperature. *Geophysical Research Letters,* **32**, L14703, doi:10.1029/2005GL022760.

Trenberth, K.E., G.W. Branstator, D. Karoly, A. Kumar, N.-C. Lau, and C. Ropelewski, 1998: Progress during TOGA in understanding and modeling global teleconnections associated with tropical sea surface temperature. *Journal of Geophysical Research,* **103(C7)**, 14291-14324.

Trenberth, K.E., J.M. Caron, and D.P. Stepaniak, 2001: The atmospheric energy budget and implications for surface fluxes and ocean heat transports. *Climate Dynamics,* **17(4)**, 259-276.

Trenberth, K.E., D.P. Stepaniak, and J.M. Caron, 2002a: Interannual variations in the atmospheric heat budget. *Journal of Geophysical Research,* **107(D8)**, 4066, doi:10.1029/2000JD000297.

Trenberth, K.E., T.R. Karl, and T.W. Spence, 2002b: The need for a systems approach to climate observations. *Bulletin of the American Meteorological Society,* **83(11),** 1593-1602.

Trenberth, K.E., J. Fasullo, and L. Smith, 2005: Trends and variability in column-integrated atmospheric water vapor. *Climate Dynamics,* **24(7-8),** 741-758.

Tribbia, J., M. Rienecker, E. Harrison, T. Rosati, and M. Ji, 2003: *Report of the "Coupled Data Assimilation Workshop",* April 21-23, 2003, Portland, Oregon. 23 pp. <http://www.usclivar.org/Meeting_Files/SSC-11/CoupledDA_rept_final.pdf>

Uppala, S.M., P.W. Kållberg, A.J. Simmons, U. Andrae, V. Da Costa Bechtold, M. Fiorino, J.K. Gibson, J. Haseler, A. Hernandez, G.A. Kelly, X. Li, K. Onogi, S. Saarinen, N. Sokka, R.P. Allan, E. Andersson, K. Arpe, M.A. Balmaseda, A.C.M. Beljaars, L. Van De Berg, J. Bidlot, N. Bormann, S. Caires, F. Chevallier, A. Dethof, M. Dragosavac, M. Fisher, M. Fuentes, S. Hagemann, E. Hólm, B.J. Hoskins, L. Isaksen, P.A.E.M. Janssen, R. Jenne, A.P. Mcnally, J.-F. Mahfouf, J.-J. Morcrette, N.A. Rayner, R.W. Saunders, P. Simon, A. Sterl, K.E. Trenberth, A. Untch, D. Vasiljevic, P. Viterbo, and J. Woollen, 2005: The ERA-40 re-analysis. *Quarterly Journal of the Royal Meteorological Society,* **131(612),** 2961-3012.

Van den Hurk, B., 2002: European LDAS established. *GEWEX Newsletter,* **12(2),** 9.

Van den Hurk, B., J. Ettema, and P. Viterbo, 2008: Analysis of soil moisture changes in Europe during a single growing season in a new ECMWF soil moisture assimilation system. *Journal of Hydrometeorology,* **9(1),** 116-131.

Wallace, J.M., 2000: North Atlantic Oscillation/Annular Mode: two paradigms–one phenomenon. *Quarterly Journal of the Royal Meteorological Society,* **126(563),** 791-805.

Wallace, J.M. and D.S. Gutzler, 1981: Teleconnections in the geopotential height field during the Northern Hemisphere winter. *Monthly Weather Review,* **109(4),** 784-812.

Walsh, J.E. and W.L. Chapman, 1998: Arctic cloud–radiation–temperature associations in observational data and atmospheric reanalyses. *Journal of Climate,* **11(11),** 3030-3045.

Wang, J., H.L. Cole, D.J. Carlson, E.R. Miller, K. Beierle, A. Paukkunen, and T.K. Laine, 2002: Corrections of humidity measurement errors from the Vaisala RS80 radiosonde: application to TOGA COARE data. *Journal of Atmospheric and Oceanic Technology,* **19(7),** 981-1002.

WCRP (World Climate Research Programme), 1998: *Proceedings of the first WCRP International Conference on Reanalyses.* Silver Spring, Maryland, USA, 27-31 October 1997. WCRP-104, WMO/TD-No. 876. World Meteorological Organization, Geneva (Switzerland), 461 pp.

WCRP (World Climate Research Programme), 2000: *Proceedings of the Second WCRP International Conference on Reanalyses.* Wokefield Park, UK, 23-27 August 1999. WCRP-109, WMO/TD-No. 985. World Meteorological Organization, Geneva (Switzerland), 452 pp.

Weaver, S.J. and S. Nigam, 2008: Variability of the Great Plains low-level jet: large-scale circulation context and hydroclimate impacts. *Journal of Climate,* **21(7),** 1532-1551.

Whitaker, J.S., G.P. Compo, X. Wei, and T.M. Hamill, 2004: Reanalysis without radiosondes using ensemble data assimilation. *Monthly Weather Review,* **132(5),** 1190-1200.

Worley, S.J., S.D. Woodruff, R.W. Reynolds, S.J. Lubker, and N. Lott, 2005: ICOADS Release 2.1 data and products. *International Journal of Climatology,* **25(7),** 823-842.

Wunsch, C., 2006: *Discrete Inverse and State Estimation Problems: With Geophysical Fluid Applications.* Cambridge University Press, Cambridge, UK, 371 pp.

Xie, P. and P.A. Arkin, 1997: Global precipitation: A 17-year monthly analysis based on gauge observations, satellite estimates and numerical model outputs. *Bulletin of the American Meteorological Society,* **78(11),** 2539-2558.

Zhang, Y., J.M. Wallace, and D.S. Battisti, 1997: ENSO-like interdecadal variability: 1900-93. *Journal of Climate,* **10(5),** 1004-1020.

Zhang, S., M.J. Harrison, A. Rosati, and A. Wittenberg, 2007: System design and evaluation of coupled ensemble data assimilation for global oceanic climate studies. *Monthly Weather Review,* **135(10),** 3541-3564.

Zipser, E.J. and R.H. Johnson, 1998: Systematic errors in radiosonde humidities a global problem? In: *Proceedings of the Tenth Symposium on Meteorological Observations and Instrumentation, January 11-16, 1998, Phoenix, Arizona.* American Meteorological Society, Boston, pp. 72-73.

Zolina, O., A. Kapala, C. Simmer, and S.K. Gulev, 2004: Analysis of extreme precipitation over Europe from different reanalyses: a comparative assessment. *Global and Planetary Change,* **44(1-4),** 129-161.

CHAPTER 3 REFERENCES

Adger, W.N., S. Agrawala, M.M.Q. Mirza, C. Conde, K. O'Brien, J. Pulhin, R. Pulwarty, B. Smit, and K. Takahashi, 2007: Assessment of adaptation practices, options, constraints and capacity. In: *Climate Change 2007: Impacts, Adaptation and Vulnerability.* Contribution of Working Group II to the Fourth Assessment Report (AR4) of the Intergovernmental Panel on Climate Change [Parry, M.L., O.F. Canziani, J.P. Palutikof, P.J.

van der Linden, and C.E. Hanson (eds.)]. Cambridge University Press, Cambridge, UK, and New York, pp. 717-743.

Alexander, M.A., I. Bladé, M. Newman, J.R. Lanzante, N.-C. Lau, and J.D. Scott, 2002: The atmospheric bridge: the influence of ENSO teleconnections on air-sea interaction over the global oceans. *Journal of Climate*, **15(16)**, 2205-2231.

Allen, M.R. and S.F.B. Tett, 1999: Checking for model inconsistency in optimal fingerprinting. *Climate Dynamics*, **15(6)**, 419-434.

Alley, R.B., J. Marotzke, W.D. Nordhaus, J.T. Overpeck, D.M. Peteet, R.A. Pielke Jr., R.T. Pierrehumbert, P.B. Rhines, T.F. Stocker, L.D. Talley, and J.M. Wallace, 2003: Abrupt climate change. *Science*, **299(5615)**, 2005-2010.

AMS (American Meteorological Society), 2004: AMS statement on meteorological drought. *Bulletin of the American Meteorological Society*, **85(5)**, 771-773. <http://www.ametsoc.org/policy/droughtstatementfinal0304.html>.

Andreadis, K.M. and D.P. Lettenmaier, 2006: Trends in 20th century drought over the continental United States. *Geophysical Research Letters*, **33**, L10403, doi:10.1029/2006GL025711.

Atlas, R., N. Wolfson, and J. Terry, 1993: The effect of SST and soil moisture anomalies on GLA model simulations of the 1988 U. S. summer drought. *Journal of Climate*, **6(11)**, 2034-2048.

Barlow, M., S. Nigam, and E.H. Berbery, 2001: ENSO, Pacific decadal variability, and U. S. summertime precipitation, drought and stream flow. *Journal of Climate*, **14(9)**, 2105-2128.

Barsugli, J.J. and P.D. Sardeshmukh, 2002: Global atmospheric sensitivity to tropical SST anomalies throughout the Indo-Pacific basin. *Journal of Climate*, **15(23)**, 3427-3442.

Bates, G.T., M.P. Hoerling, and A. Kumar, 2001: Central U. S. springtime precipitation extremes: teleconnections and relationships with sea surface temperature. *Journal of Climate*, **14(17)**, 3751-3766.

Bjerknes, J., 1966: A possible response of the atmospheric Hadley circulation to equatorial anomalies of ocean temperature. *Tellus*, **18(4)**, 820-828.

Bjerknes, J., 1969: Atmospheric teleconnections from the equatorial Pacific. *Monthly Weather Review*, **97(3)**, 163-172.

Bracco, A., F. Kucharski, F. Molteni, W. Hazeleger, and C. Severjins, 2007: A recipe for simulating the interannual variability of the Asian summer monsoon and its relation with ENSO. *Climate Dynamics*, **28(5)**, 441-460.

Breshears, D.D., N.S. Cobb, P.M. Rich, K.P. Price, C.D. Allen, R.G. Balice, W.H. Romme, J.H. Kastens, M.L. Floyd, J. Belnap, J.J. Anderson, O.B. Myers, and C.W. Meyer, 2005: Regional vegetation die-off in response to global-change type drought. *Proceedings of the National Academy of Sciences*, **102(42)**, 15144-15148.

Broecker, W.S., 1975: Climatic change: Are we on the brink of a pronounced global warming? *Science*, **189(4201)**, 460-463.

Broecker, W.S., 2003: Does the trigger for abrupt climate change reside in the ocean or in the atmosphere? *Science*, **300(5625)**, 1519-1522.

CCSP (Climate Change Science Program), 2008: *Weather and Climate Extremes in a Changing Climate: Regions of Focus: North America, Hawaii, Caribbean, and U.S. Pacific Islands.* [Karl, T.R., G.A. Meehl, C.D. Miller, S.J. Hassol, A.M. Waple, and W.L. Murray (eds.)]. Synthesis and Assessment Product 3.3. U.S. Climate Change Science Program, Washington, DC, 164 pp.

Chen, P. and M. Newman, 1998: Rossby wave propagation and the rapid development of upper-level anomalous anticyclones during the 1988 U.S. drought. *Journal of Climate*, **11(10)**, 2491-2504.

Christy, J.R., W.B. Norris, K. Redmond, and K.P. Gallo, 2006: Methodology and results of calculating central California surface temperature trends: evidence of human-caused climate change? *Journal of Climate*, **19(4)**, 548-563.

Church, J.A., N.J. White, and J.M. Arblaster, 2005: Significant decadal-scale impact of volcanic eruptions on sea level and ocean heat content. *Nature*, **438(7064)**, 74-77.

Clarke, P., N. Pisias, T. Stocker, and A. Weaver, 2002: The role of the thermohaline circulation in abrupt climate change. *Nature*, **415(6874)**, 863-869.

Clement, A.C., R. Seager, M.A. Cane, and S.E. Zebiak, 1996: An ocean dynamical thermostat. *Journal of Climate*, **9(9)**, 2190-2196.

Cole, J.E., J.T. Overpeck, and E.R. Cook, 2002: Multiyear La Niña events and persistent drought in the contiguous United States. *Geophysical Research Letters*, **29(13)**, 1647, doi:10.1029/2001GL013561.

Cook, E.R., C.A. Woodhouse, C.M. Eakin, D.M. Meko, and D.W. Stahle, 2004: Long-term aridity changes in the western United States. *Science*, **306(5698)**, 1015-1018.

Dai, A. and K.E. Trenberth, 2004: The diurnal cycle and its depiction in the Community Climate System Model. *Journal of Climate*, **17(5)**, 930-951.

Dai, A., K.E. Trenberth, and T.T. Qian, 2004: A global dataset of Palmer Drought Severity Index for 1870-2002: relationship with soil moisture and effects of surface warming. *Journal of Hydrometeorology*, **5(6)**, 1117-1130.

Delworth, T.L. and T.R. Knutson, 2000: Simulation of early 20th century global warming. *Science*, **287(5461)**, 2246-2250.

Diaz, H.F., 1983: Some aspects of major dry and wet periods in the contiguous United States 1895-1981. *Journal of Climate and Applied Meteorology*, **22(1)**, 3-6.

Feldstein, S.B., 2000: The timescale, power spectra, and climate noise properties of teleconnection patterns. *Journal of Climate*, **13(24)**, 4430-4440.

Feldstein, S., 2002: Fundamental mechanisms of the growth and decay of the PNA teleconnection pattern. *Quarterly Journal of the Royal Meteorological Society*, **128(581)**, 775-796.

Gates, W.L., 1992: AMIP: The Atmospheric Model Intercomparison Project. *Bulletin of the American Meteorological Society*, **73(12)**, 1962-1970.

Gillett, N.P., F.W. Zwiers, A.J. Weaver, and P.A. Stott, 2003: Detection of human influence on sea-level pressure. *Nature*, **422(6929)**, 292-294.

Gillett, N.P., A.J. Weaver, F.W. Zwiers, and M.F. Wehner, 2004: Detection of volcanic influence on global precipitation. *Geophysical Research Letters*, **31**(12), L12217, doi:10.1029/2004GL020044.

Glantz, M.H., R.W. Katz, and N. Nicholls (eds.), 1991: *Teleconnections Linking Worldwide Climate Anomalies: Scientific Basis and Societal Impact.* Cambridge University Press, Cambridge, UK, and New York, 535 pp.

Groisman, P.Ya., R.W. Knight, T.R. Karl, D.R. Easterling, B. Sun, and J.H. Lawrimore, 2004: Contemporary changes of the hydrological cycle over the contiguous United States: trends derived from *in situ* observations. *Journal of Hydrometeorology*, **5(1)**, 64-85.

Hale, R.C., K.P. Gallo, T.W. Owen, and T.R. Loveland, 2006: Land use/land cover change effects on temperature trends at U.S. climate normals stations. *Geophysical Research Letters*, **33**, L11703, doi:10.1029/2006GL026358.

Hasselmann, K., 1979: On the signal-to-noise problem in atmospheric response studies. In: *Meteorology Over the Tropical Oceans* [Shaw, D.B. (ed.)]. Royal Meteorological Society, Bracknell (UK), pp. 251-259.

Hasselmann, K., 1997: Multi-pattern fingerprint method for detection and attribution of climate change. *Climate Dynamics*, **13(9)**, 601-612.

Hegerl, G.C., F.W. Zwiers, P. Braconnot, N.P. Gillett, Y. Luo, J. Marengo, N. Nicholls, J.E. Penner, and P.A. Stott, 2007: Understanding and attributing climate change. In: *Climate Change 2007: The Physical Science Basis.* Contribution of Working Group I to the Fourth Assessment Report (AR4) of the Intergovernmental Panel on Climate Change [Solomon, S., D. Qin, M. Manning, Z. Chen, M. Marquis, K.B. Averyt, M. Tignor, and H.L. Miller (eds.)]. Cambridge University Press, Cambridge, UK, and New York, pp. 663-745.

Herweijer, C., R. Seager, and E.R. Cook, 2006: North American droughts of the mid to late nineteenth century: a history, simulation and implication for mediaeval drought. *Holocene*, **16(2)**, 159-171.

Higgins, R.W., Y. Chen, and A.V. Douglas, 1999: Interannual variability of the North American warm season precipitation regime. *Journal of Climate*, **12(3)**, 653-680.

Hoerling, M.P. and A. Kumar, 2002: Atmospheric response patterns associated with tropical forcing. *Journal of Climate*, **15(16)**, 2184-2203.

Hoerling, M. and A. Kumar, 2003: The perfect ocean for drought. *Science*, **299(5607)**, 691-694.

Hoerling, M.P., J.W. Hurrell, and T.Y. Xu, 2001: Tropical origins for recent North Atlantic climate change. *Science*, **292(5514)**, 90-92.

Hoerling, M.P., J.W. Hurrell, T. Xu, G.T. Bates, and A. Phillips, 2004: Twentieth century North Atlantic climate change. Part II: understanding the effect of Indian Ocean warming. *Climate Dynamics*, **23(3-4)**, 391-405.

Hoerling, M., J. Eischeid, X. Quan, and T. Xu, 2007: Explaining the record US warmth of 2006. *Geophysical Research Letters*, **34**, L17704, doi:10.1029/2007GL030643.

Hong, S-Y. and E. Kalnay, 2002: The 1998 Oklahoma-Texas drought: mechanistic experiments with NCEP global and regional models. *Journal of Climate*, **15(9)**, 945-963.

Horel, J.D. and J.M. Wallace, 1981: Planetary-scale atmospheric phenomena associated with the Southern Oscillation. *Monthly Weather Review*, **109(4)**, 813-829.

Hoskins, B.J. and D.J. Karoly, 1981: The steady linear response of a spherical atmosphere to thermal and orographic forcing. *Journal of the Atmospheric Sciences*, **38(6)**, 1179-1196.

Houghton, J.T., L.G. Meira Filho, B.A. Callander, N. Harris, A. Kattenberg, and K. Maksell (eds.), 1996: *Climate Change 1995: The Science of Climate Change.* Cambridge University Press, Cambridge, UK, and New York, 572 pp.

Hurrell, J.W., 1995: Decadal trends in the North-Atlantic Oscillation: regional temperatures and precipitation. *Science*, **269(5224)**, 676-679.

Hurrell, J.W., 1996: Influence of variations in extratropical wintertime teleconnections on Northern Hemisphere temperatures. *Geophysical Research Letters*, **23(6)**, 665-668.

Huschke, R.E. (ed.), 1959: *Glossary of Meteorology.* Boston, American Meteorological Society, 638 pp.

IDAG (International Ad Hoc Detection and Attribution Group), 2005: Detecting and attributing external influences on the climate system: A review of recent advances. *Journal of Climate,* **18(9),** 1291-1314.

IPCC (International Panel on Climate Change), 2001: *Climate Change 2001: The Scientific Basis.* Contribution of Working Group I to the Third Assessment Report of the Intergovernmental Panel on Climate Change [Houghton, J.T., Y. Ding, D.J. Griggs, M. Noguer, P.J. van der Linden, X. Dai, K. Maskell, and C.A. Johnson (eds.)]. Cambridge University Press, Cambridge, UK, and New York, 881 pp.

IPCC (Intergovernmental Panel on Climate Change) 2007a: *Climate Change 2007: The Physical Science Basis.* Contribution of Working Group I to the Fourth Assessment Report (AR4) of the Intergovernmental Panel on Climate Change [Solomon, S., D. Qin, M. Manning, Z. Chen, M. Marquis, K.B. Averyt, M. Tignor, and H.L. Miller (eds.)]. Cambridge University Press, Cambridge, UK, and New York, 996 pp. <http://www.ipcc.ch>

IPCC (Intergovernmental Panel on Climate Change) 2007b: Summary for Policy Makers. In: *Climate Change 2007: Impacts, Adaptation and Vulnerability.* Contribution of Working Group II to the Fourth Assessment Report (AR4) of the Intergovernmental Panel on Climate Change [Parry, M.L., O.F. Canziani, J.P. Palutikof, P.J. van der Linden, and C.E. Hanson (eds.)]. Cambridge University Press, Cambridge, UK, and New York, pp. 7-22.

Kalnay, E. and M. Cai, 2003: Impact of urbanization and land-use on climate change. *Nature,* **423(6939),** 528-531.

Kalnay, E., M. Kanamitsu, R. Kistler, W. Collins, D. Deaven, L. Gandin, M. Iredell, S. Saha, G. White, J. Woollen, Y. Zhu, A. Leetmaa, R. Reynolds, M. Chelliah, W. Ebisuzaki, W. Higgins, J. Janowiak, K. C. Mo, C. Ropelewski, J. Wang, R. Jenne, and D. Joseph, 1996: The NCEP/NCAR 40-Year Reanalysis Project. *Bulletin of the American Meteorological Society,* **77(3),** 437-471.

Kalnay, E., M. Cai, H. Li, and J. Tobin, 2006: Estimation of the impact of land-surface forcings on temperature trends in eastern Unites States. *Journal of Geophysical Research,* **111,** D06106, doi:10.1029/2005JD006555.

Karl, T.R., 1983: Some spatial characteristics of drought duration in the United States. *Journal of Climate and Applied Meteorology,* **22(8),** 1356-1366.

Karoly, D.J. and Q. Wu, 2005: Detection of regional surface temperature trends. *Journal of Climate,* **18(21),** 4337-4343.

Karoly, D.J., K. Braganza, P.A. Stott, J. Arblaster, G. Meehl, A. Broccoli, and K.W. Dixon, 2003: Detection of a human influence on North American climate. *Science* **302(5648),** 1200-1203.

Kirchner, I, G. Stenchikov, H.-F. Graf, A. Robock, and J. Antuña, 1999: Climate model simulation of winter warming and summer cooling following the 1991 Mount Pinatubo volcanic eruption. *Journal of Geophysical Research,* **104(D16),** 19039-19055.

Klein, S.A., B.J. Sodden, and N.-C. Lau, 1999: Remote sea surface variations during ENSO: Evidence for a tropical atmospheric bridge. *Journal of Climate,* **12(4),** 917-932.

Knutson, T.R., T.L. Delworth, K.W. Dixon, I.M. Held, J. Lu, V. Ramaswamy, M.D. Schwarzkopf, G. Stenchikov, and R.J. Stouffer, 2006: Assessment of twentieth-century regional surface temperature trends using the GFDL CM2 coupled models. *Journal of Climate,* **19(9),** 1624-1651.

Kueppers, L.M., M.A. Snyder, L.C. Sloan, D. Cayan, J. Jin, H. Kanamaru, M. Kanamitsu, N.L. Miller, M. Tyree, H. Du, and B. Weare, 2007: Seasonal temperature responses to land use change in the western United States. *Global and Planetary Change,* **60(3-4),** 250-264.

Kukla, G.J. and R.K. Matthews, 1972: When will the present interglacial end? *Science,* **178(4057),** 190-191.

Kumar, A., W. Wang, M.P. Hoerling, A. Leetmaa, and M. Ji, 2001: The sustained North American warming of 1997 and 1998. *Journal of Climate,* **14(3),** 345-353.

Kunkel, K.E., X.-Z. Liang, J. Zhu, and Y. Lin, 2006: Can CGCMS simulate the twentieth century "warming hole" in the central United States? *Journal of Climate,* **19(17),** 4137-4153.

Kushnir, Y., W.A. Robinson, I. Bladé, N.M.J. Hall, S. Peng, and R. Sutton, 2002: Atmospheric GCM response to extratropical SST anomalies: synthesis and evaluation. *Journal of Climate,* **15(16),** 2233-2256.

Lambert, F.H., P.A. Stott, M.R. Allen, and M.A. Palmer, 2004: Detection and attribution of changes in 20th century land precipitation. *Geophysical Research Letters,* **31(10),** L10203, doi:10.1029/2004GL019545.

Latif, M. and T.P. Barnett, 1996: Decadal climate variability over the North Pacific and North America: dynamics and predictability. *Journal of Climate,* **9(10),** 2407-2423.

Lau, N.-C and M.J. Nath, 2003: Atmosphere-ocean variations in the Indo-Pacific sector during ENSO episodes. *Journal of Climate,* **16(1),** 3-20.

Lau, N.-C., A. Leetmaa, M.J. Nath, and H.-L. Wang, 2005: Influences of ENSO-induced Indo-Western Pacific SST anomalies on extratropical atmospheric variability during the boreal summer. *Journal of Climate*, **18(15)**, 2922-2942.

Lau, N.-C., A. Leetmaa, and M.J. Nath, 2006: Attribution of atmospheric variations in the 1997-2003 period to SST anomalies in the Pacific and Indian Ocean basins. *Journal of Climate*, **19(15)**, 3607-3628.

Liu, A.Z., M. Ting, and H. Wang, 1998: Maintenance of circulation anomalies during the 1988 drought and 1993 floods over the United States. *Journal of the Atmospheric Sciences*, **55(17)**, 2810-2832.

Livezey, R.E., M. Masutani, A. Leetmaa, H.L. Rui, M. Ji, and A. Kumar, 1997: Teleconnective response of the Pacific-North American region atmosphere to large central equatorial Pacific SST anomalies. *Journal of Climate*, **10(8)**, 1787-1820.

Lyon, B. and R.M. Dole, 1995: A diagnostic comparison of the 1980 and 1988 U. S. summer heat wave-droughts. *Journal of Climate*, **8(6)**, 1658-1675.

Mahmood, R., S.A. Foster, T. Keeling, K.G. Hubbard, C. Carlson, and R. Leeper, 2006: Impacts of irrigation on 20th century temperature in the northern Great Plains. *Global and Planetary Change*, **54(1-2)**, 1-18.

Mantua, N.J., S.R. Hare, Y. Zhang, J.M. Wallace, and R.C. Francis, 1997: A Pacific inter-decadal climate oscillation with impacts on salmon production. *Bulletin of the American Meteorological Society*, **78(6)**, 1069-1079.

McCabe, G.J., M.A. Palecki, and J.L. Betencourt, 2004: Pacific and Atlantic Ocean influences on multidecadal drought frequency in the United States. *Proceedings of the National Academy of Sciences*, **101(12)**, 4136-4141.

McPherson, R.A., 2007: A review of vegetation–atmosphere interactions and their influences on mesoscale phenomena. *Progress in Physical Geography*, **31(3)**, 261-285.

Mendelssohn, R., S.J. Bograd, F.B. Schwing, and D.M. Palacios, 2005: Teaching old indices new tricks: A state-space analysis of El Niño related climate indices. *Geophysical Research Letters*, **32**, L07709, doi:10.1029/2005GL022350.

Merryfield, W.J., 2006: Changes to ENSO under CO_2 doubling in a multimodal ensemble. *Journal of Climate*, **19(16)**, 4009-4027.

Miller, A.J., D.R. Cayan, T.P. Barnett, N.E. Graham, and J.M. Oberhuber, 1994: The 1976-77 climate shift of the Pacific Ocean. *Oceanography*, **7(1)**, 21-26.

Mitchell, J.F.B., D.J. Karoly, G.C. Hegerl, F.W. Zwiers, M.R. Allen, and J. Marengo, 2001: Detection of climate change and attribution of causes. In: *Climate Change 2001: The Scientific Basis*. Contribution of Working Group I to the Third Assessment Report of the Intergovernmental Panel on Climate Change [Houghton, J.T., Y. Ding, D.J. Griggs, M. Noguer, P.J. van der Linden, X. Dai, K. Maskell, and C.A. Johnson (eds.)]. Cambridge University Press, Cambridge, UK, and New York, pp. 695-738.

Namias, J., 1966: Nature and possible causes of the northeastern United States drought during 1962-1965. *Monthly Weather Review*, **94(9)**, 543-554.

Namias, J., 1978: Multiple causes of the North American abnormal winter 1976-77. *Monthly Weather Review*, **106(3)**, 279-295.

Namias, J., 1983: Some causes of United States drought. *Journal of Climate and Applied Meteorology*, **22(1)**, 30-39.

Namias, J., 1991: Spring and summer 1988 drought over the contiguous United States – causes and prediction. *Journal of Climate*, **4(1)**, 54-65.

Narisma, G., J. Foley, R. Licker, and N. Ramankutty, 2007: Abrupt changes in rainfall during the twentieth century. *Geophysical Research Letters*, **34**, L06710, doi:10.1029/2006GL028628.

NIDIS (National Integrated Drought Information System), 2004: *Creating a Drought Early Warning System for the 21st Century: The National Integrated Drought Information System (NIDIS)*. Western Governors' Association, Denver, CO, 13 pp. <http://www.westgov.org/wga/initiatives/drought/>

Newman, M. and P.D. Sardeshmukh, 1998: The impact of the annual cycle on the North Pacific/North American response to remote low-frequency forcing. *Journal of the Atmospheric Sciences*, **55(8)**, 1336-1353.

NRC (National Research Council), 1975: *Understanding Climatic Change: A Program For Action*. National Academy of Sciences, Washington DC, 239 pp.

NRC (National Research Council), 2002: *Abrupt Climate Change: Inevitable Surprises*. National Academy Press, Washington DC, 230 pp.

Palmer, W.C., 1965: *Meteorological drought*. Research paper no. 45. U.S. Department of Commerce, Weather Bureau, Washington, DC, 58 pp.

Pielke, R.A., Sr., G. Marland, R.A. Betts, T.N. Chase, J.L. Eastman, J.O. Niles, D.S. Niyogi, and S.W. Running, 2002: The influence of land-use change and landscape dynamics on the climate system: relevance to climate-change policy beyond the radiative effect of greenhouse gases. *Philosophical Transactions of the Royal Society of London Series A*, **360(1797)**, 1705-1719.

Rasmusson, E.M. and J.M. Wallace, 1983: Meteorological aspects of the El Niño/Southern Oscillation. *Science*, **222(4629)**, 1195-1202.

Robertson, A.W., J.D. Farrara, and C.R. Mechoso, 2003: Simulations of the atmospheric response to south Atlantic sea surface temperature anomalies. *Journal of Climate*, **16(15)**, 2540-2551.

Robinson, W.A., R. Reudy, and J.E. Hansen, 2002: General circulation model simulations of recent cooling in the east-central United States. *Journal of Geophysical Research*, **107(D24)**, 4748, doi:10.1029/2001JD001577.

Ropelewski, C.F. and M.S. Halpert, 1986: North American precipitation and temperature patterns associated with the El Niño/Southern Oscillation (ENSO). *Monthly Weather Review*, **114(12)**, 2352-2362.

Rossby, C.G., 1939: Relation between variations in the intensity of the zonal circulation of the atmosphere and the displacements of the semi-permanent centers of action. *Journal of Marine Research*, **2(1)**, 38-55.

Santer, B.D., M.F. Wehner, T.M.L. Wigley, R. Sausen, G.A. Meehl, K.E. Taylor, C. Ammann, J. Arblaster, W.M. Washington, J.S. Boyle, and W. Brüggemann, 2003: Contributions of anthropogenic and natural forcing to recent tropopause height changes. *Science*, **301(5632)**, 479-483.

Santer, B.D., T.M.L. Wigley, P.J. Gleckler, C. Bonfils, M.F. Wehner, K. AchutaRao, T.P. Barnett, J.S. Boyle, W. Brüggemann, M. Fiorino, N. Gillett, J.E. Hansen, P.D. Jones, S.A. Klein, G.A. Meehl, S.C.B. Raper, R.W. Reynolds, K.E. Taylor, and W.M. Washington, 2006: Forced and unforced ocean temperature changes in the Atlantic and Pacific tropical cyclogenesis regions. *Proceedings of the National Academy of Sciences*, **103(38)**, 13905-13910.

Schubert, S.D., M.J. Suarez, P.J. Pegion, R.D. Koster, and J.T. Bacmeister, 2004: Causes of long-term drought in the U.S. Great Plains. *Journal of Climate*, **17(3)**, 485-503.

Seager, R., Y. Kushnir, C. Herweijer, N. Naik, and J. Velez, 2005: Modeling of tropical forcing of persistent droughts and pluvials over western North America: 1856-2000. *Journal of Climate*, **18(19)**, 4065-4088.

Seager, R., M. Ting, I. Held, Y. Kushnir, J. Lu, G. Vecchi, H.-P. Huang, N. Harnik, A. Leetmaa, N.-C. Lau, C. Li, J. Velez, and N. Naik, 2007: Model projections of an imminent transition to a more arid climate in southwestern North America. *Science*, **316(5828)**, 1181-1184.

Simmons, A.J., J.M. Wallace, and G. Branstator, 1983: Barotropic wave propagation and instability and atmospheric teleconnection patterns. *Journal of the Atmospheric Sciences*, **40(6)**, 1363-1392.

Smith, J.B., H.-J. Schellnhuber, and M.M.Q. Mirza, 2001: Vulnerability to climate change and reasons for concern: a synthesis. In: *Climate Change 2001: Impacts, Adaptation, and Vulnerability*. Contribution of Working Group II to the Report of the Intergovernmental Panel on Climate Change [McCarthy, J.J., O.F. Canziani, M.A. Leary, D.J. Dokken, and K.S. White (eds)]. Cambridge University Press, Cambridge, UK, and New York, pp. 913-967.

Soja, A.J., N.M. Tchebakova, N.H.F. French, M.D. Flannigan, H.H. Shugart, B.J. Stocks, A.I. Sukhinin, E.I. Parfenova, F.S. Chapin III, and P.W. Stackhouse Jr., 2007: Climate-induced boreal forest change: predictions versus current observations. *Global and Planetary Change*, **56(3-4)**, 274-296.

Stahle, D.W., M.K. Cleaveland, M.D. Therrell, D.A. Gay, R.D. D'Arrigo, P.J. Krusic, E.R. Cook, R.J. Allan, J.E. Cole, R.B. Dunbar, M.D. Moore, M.A. Stokes, B.T. Burns, J. Villanueva-Diaz, and L.G. Thompson, 1998: Experimental dendroclimatic reconstruction of the Southern Oscillation. *Bulletin of the American Meteorological Society*, **79(10)**, 2137-2152.

Stott, P.A., 2003: Attribution of regional-scale temperature changes to anthropogenic and natural causes. *Geophysical Research Letters*, **30(14)**, 1724, doi:10.1029/2003GL017324.

Stott, P.A. and S.F.B. Tett, 1998: Scale-dependent detection of climate change. *Journal of Climate*, **11(12)**, 3282-3294.

Therrell, M.D., D.W. Stahle, M.K. Cleaveland, and J. Villanueva-Diaz, 2002: Warm season tree growth and precipitation over Mexico. *Journal of Geophysical Research*, **107(D14)**, 4205, doi:10.1029/2001JD000851.

Thompson, D.W.J. and J.M. Wallace, 1998: The Arctic Oscillation signature in the wintertime geopotential height temperature fields. *Geophysical Research Letters*, **25(9)**, 1297-1300.

Thompson, D.W.J. and J.M. Wallace, 2000a: Annular modes in the extratropical circulation. Part I: month-to-month variability. *Journal of Climate*, **13(5)**, 1000-1016.

Thompson, D.W.J. and J.M. Wallace, 2000b: Annular modes in the extratropical circulation. Part II: trends. *Journal of Climate*, **13(5)**, 1018-1036.

Ting, M. and H. Wang, 1997: Summertime U.S. precipitation variability and its relation to Pacific sea surface temperature. *Journal of Climate*, **10(8)**, 1853-1873.

Trenberth, K.E., 1990: Recent observed interdecadal climate changes in the Northern Hemisphere. *Bulletin of the American Meteorological Society*, **71(7)**, 988-993.

Trenberth, K.E., 2004: Rural land-use change and climate. *Nature*, **427(6971)**, 213.

Trenberth, K.E. and G.W. Branstator, 1992: Issues in establishing causes of the 1988 drought over North America. *Journal of Climate*, **5(2)**, 159-172.

Trenberth, K. and T.J. Hoar, 1996: The 1990-1995 El Niño-Southern Oscillation event: longest on record. *Geophysical Research Letters*, **23**, 57-60.

Trenberth, K.E., G.W. Branstator, and P.A. Arkin, 1988: Origins of the 1988 North American drought. *Science*, **242(4886)**, 1640-1645.

Trenberth, K.E., G.W. Branstrator, D. Karoly, A. Kumar, N.-C. Lau, and C. Ropelewski, 1998: Progress during TOGA in understanding and modeling global teleconnections associated with tropical sea surface temperatures. *Journal of Geophysical Research*, **103(C7)**, 14291-14324.

Vecchi, G.A. and B. Soden, 2007: Global warming and the weakening of the tropical circulation. *Journal of Climate*, **20(17)**, 4316-4340.

Vose, R.S., T.R. Karl, D.R. Easterling, C.N. Williams, and M.J. Menne, 2004: Impact of land-use change on climate. *Nature*, **427(6971)**, 213-214.

Walker, G.T. and E.W. Bliss, 1932: World weather V. *Memoirs of the Royal Meteorological Society*, **4(36)**, 53-84.

Wallace, J.M. and D.S. Gutzler, 1981: Teleconnections in the geopotential height field during the Northern Hemisphere winter. *Monthly Weather Review*, **109(4)**, 784-812.

Wolfson, N., R. Atlas, and Y.C. Sud, 1987: Numerical experiments related to the summer 1980 U.S. heat wave. *Monthly Weather Review*, **115(7)**, 1345-1357.

Wu, Q. and D.J. Karoly, 2007: Implications of changes in the atmospheric circulation on the detection of regional surface air temperature trends. *Geophysical Research Letters*, **34**, L08703, doi:10.1029/2006GL028502.

Yu, L. and M.M. Rienecker, 1999: Mechanisms for the Indian Ocean warming during the 1997-1998 El Niño. *Geophysical Research Letters*, **26(6)**, 735-738.

Zhang, X., F.W. Zwiers, and P.A. Stott, 2006: Multimodel multisignal climate change detection at regional scale. *Journal of Climate*, **19(17)**, 4294-4307.

Zhang, X., F.W. Zwiers, G.C. Hegerl, F.H. Lambert, N. P. Gillett, S. Solomon, P.A. Stott, and T. Nozawa, 2007: Detection of human influence on twentieth-century precipitation trends. *Nature*, **448(7152)**, 461-465.

Zwiers, F.W. and X. Zhang, 2003: Towards regional scale climate change detection. *Journal of Climate,* **16(5)**, 793-797.

CHAPTER 4 REFERENCES

Allen, M.R., 2003: Liability for climate change. *Nature,* **421(6926)**, 891-892.

Allen, M.R., N.P. Gillett, J.A. Kettleborough, G. Hegerl, R. Schnur, P.A. Stott, G. Boer, C. Covey, T.L. Delworth, G.S. Jones, J.F.B. Mitchell, and T.P. Barnett, 2006: Quantifying anthropogenic influence on recent near-surface temperature change. *Surveys in Geophysics,* **27(5)**, 491-544.

Arkin, P., E. Kalnay, J. Laver, S. Schubert, and K. Trenberth, 2004: *Ongoing Analysis of the Climate System: A Workshop Report.* Boulder, CO, 18-20 August 2003. University Corporation for Atmospheric Research, Boulder, CO, 48 pp. <http://www.joss.ucar.edu/joss_psg/meetings/archived/climatesystem/index.html>

Baidya R.S. and R. Avissar, 2002: Impact of land use/land cover change on regional hydrometeorology in the Amazon. *Journal of Geophysical Research*, **107(D20)**, 8037, doi:10.1029/2000JD000266.

Bengtsson, L., P. Arkin, P. Berrisford, P. Bougeault, C.K. Folland, C. Gordon, K. Haines, K.I. Hodges, P. Jones, P. Kållberg, N. Rayner, A.J. Simmons, D. Stammer, P.W. Thorne, S. Uppala, and R.S. Vose, 2007: The need for a dynamical climate reanalysis. *Bulletin of the American Meteorological Society,* **88(4)**, 495-501.

CCSP (Climate Change Science Program), 2003: *Strategic Plan for the U.S. Climate Change Science Program.* U.S. Climate Change Science Program, Washington, DC, 202 pp.

CCSP (Climate Change Science Program), 2006: *Temperature Trends in the Lower Atmosphere: Steps for Understanding and Reconciling Differences.* [Karl, T.R., S.J. Hassol, C.D. Miller, and W.L. Murray (eds.)]. Synthesis and Assessment Product 1.1. U.S. Climate Change Science Program, Washington, DC, 164 pp.

CCSP (Climate Change Science Program), 2008: *Weather and Climate Extremes in a Changing Climate: Regions of Focus: North America, Hawaii, Caribbean, and U.S. Pacific Islands.* [Karl, T.R., G.A. Meehl, C.D. Miller, S.J. Hassol, A.M. Waple, and W.L. Murray (eds.)]. Synthesis and Assessment Product 3.3. U.S. Climate Change Science Program, Washington, DC, 164 pp.

Challinor, A.J., T.R. Wheeler, J.M. Slingo, P.Q. Crauford, and D. Grimes, 2005: Simulation of crop yields using ERA-40: Limits to skill and nonstationarity in weather-yield relationships. *Journal of Applied Meteorology*, **44(4)**, 516-531.

Chase, T.N., R.A. Pielke, T. Kittel, R. Nemani, and S.W. Running, 2000: Simulated impacts of historical land cover changes on global climate in northern winter. *Climate Dynamics*, **16(2-3)**, 93-105.

Compo, G.P., J.S. Whitaker, and P.D. Sardeshmukh, 2006: Feasibility of a 100-year reanalysis using only surface pressure data. *Bulletin of the American Meteorological Society*, **87(2)**, 175-190.

Dee, D.P., 2005: Bias and data assimilation. *Quarterly Journal of the Royal Meteorological Society,* **131(613)**, 3323-3343.

GCOS (Global Climate Observing System), 2003: *The Second Report on the Adequacy of the Global Observing System for Climate in Support of the UNFCCC.* GCOS-82, WMO/TD 1143. GCOS Secretariat, Geneva, (Switzerland), 84 pp.

GCOS (Global Climate Observing System), 2004: *Implementation Plan for the Global Observing System for Climate in Support of the UNFCCC.* GCOS-92, WMO/TD 1219. GCOS Secretariat, Geneva, (Switzerland), 136 pp.

GEOSS, 2005: *Global Earth Observation System of Systems, GEOSS: 10-Year Implementation Plan Reference Document.* GEO1000R/ESA SP-1284. ESA Publications, Noordwijk, the Netherlands, 209 pp.

Gillett, N.P., F.W. Zwiers, A.J. Weaver, G.C. Hegerl, M.R. Allen, and P.A. Stott, 2002: Detecting anthropogenic influence with a multi-model ensemble. *Geophysical Research Letters*, **29(10)**, 1970, doi:10.1029/2002GL015836.

Hasselmann, K., 1997: Multi-pattern fingerprint method for detection and attribution of climate change. *Climate Dynamics*, **13(9)**, 601-612.

Hegerl, G.C., P.A. Stott, M.R. Allen, J.F.B. Mitchell, S.F.B Tett, and U. Cubasch, 2000: Optimal detection and attribution of climate change: sensitivity of results to climate model differences. *Climate Dynamics*, **16(10-11)**, 737-754.

Hegerl, G.C., T.R. Karl, M. Allen, N.L. Bindoff, N. Gillett, D. Karoly, X. Zhang, and F. Zwiers, 2006: Climate change detection and attribution: beyond mean temperature signals. *Journal of Climate,* **19(20)**, 5058-5077.

Hoerling, M., J. Eischeid, X. Quan, and T. Xu, 2007: Explaining the record US warmth of 2006. *Geophysical Research Letters*, **34**, L17704, doi:10.1029/2007GL030643.

Hollingsworth, A., S. Uppala, E. Klinker, D. Burridge, F. Vitart, J. Onvlee, J.W. De Vries, A. De Roo, and C. Pfrang, 2005: The transformation of earth-system observations into information of socio-economic value in GEOSS. *Quarterly Journal of the Royal Meteorological Society,* **131(613)**, 3493-3512.

Houtekamer, P.L. and H. L. Mitchell, 1998: Data assimilation using an ensemble Kalman filter technique. *Monthly Weather Review*, **126(3)**, 796-811.

IPCC (Intergovernmental Panel on Climate Change), 2007: *Climate Change 2007: The Physical Science Basis*. Contribution of Working Group I to the Fourth Assessment Report (AR4) of the Intergovernmental Panel on Climate Change [Solomon, S., D. Qin, M. Manning, Z. Chen, M. Marquis, K.B. Averyt, M. Tignor, and H.L. Miller (eds.)]. Cambridge University Press, Cambridge, UK, and New York, 996 pp. <http://www.ipcc.ch>

Kalnay, E., M. Kanamitsu, R. Kistler, W. Collins, D. Deaven, L. Gandin, M. Iredell, S. Saha, G. White, J. Woollen, Y. Zhu, A. Leetmaa, B. Reynolds, M. Chelliah, W. Ebisuzaki, W. Higgins, J. Janowiak, K.C. Mo, C. Ropelewski, J. Wang, R. Jenne, and D. Joseph, 1996: The NCEP/NCAR 40-Year Reanalysis Project. *Bulletin of the American Meteorological Society,* **77(3)**, 437-471.

Kosatsky, T., 2005: The 2003 European heat waves. *EuroSurveillance*, **10(7-8)**, 148-149.

Kunkel, K.E., X.Z. Liang, J. Zhu, and Y. Lin, 2006: Can CGCMs simulate the twentieth-century "warming hole" in the central United States? *Journal of Climate*, **19(17)**, 4137-4153.

Matthews, H.D., A.J. Weaver, K.J. Meissner, N.P. Gillett, and M. Eby, 2004: Natural and anthropogenic climate change: incorporating historical land cover change, vegetation dynamics and the global carbon cycle. *Climate Dynamics*, **22(5)**, 461-479.

Mesinger, F., G. DiMego, E. Kalnay, K. Mitchell, P.C. Shafran, W. Ebisuzaki, D. Jović, J. Woollen, E. Rogers, E.H. Berbery, M.B. Ek, Y. Fan, R. Grumbine, W. Higgins, H. Li, Y. Lin, G. Manikin, D. Parrish, and W. Shi, 2006: North American regional reanalysis. *Bulletin of the American Meteorological Society*, **87(3)**, 343-360.

NIDIS (National Integrated Drought Information System), 2007: *The National Integrated Drought Information System Implementation Plan: A Pathway for National Resilience.* NOAA Climate Program Office, Silver Spring MD, 29 pp.

Nigam, S. and A. Ruiz-Barradas, 2006: Seasonal hydroclimate variability over North America in global and regional reanalyses and AMIP simulations: varied representation. *Journal of Climate,* **19(5)**, 815-837.

NRC (National Research Council), 1991: *Four-Dimensional Model Assimilation of Data: A Strategy for the Earth System Sciences*. National Academy Press, Washington, DC, 88 pp.

Ott, E., B.R. Hunt, I. Szunyogh, A.V. Zimin, E.J. Kostelich, M. Corazza, E. Kalnay, D.J. Patil, and J.A. Yorke, 2004: A local ensemble Kalman filter for atmospheric data assimilation. *Tellus A*, **56(5)**, 415-428.

Peters W., J.B. Miller, J. Whitaker, A.S. Denning, A. Hirsch, M.C. Krol, D. Zupanski, L. Bruhwiler, and P.P. Tans, 2005: An ensemble data assimilation system to estimate CO_2 surface fluxes from atmospheric trace gas observations. *Journal of Geophysical Research*, **110**, D24304, doi:10.1029/2005JD006157.

Pielke, R.A., Sr., 2001: Influence of the spatial distribution of vegetation and soils on the prediction of cumulus convective rainfall. *Reviews of Geophysics,* **39(2)**, 151-177.

Pielke, R.A., R.L. Walko, L.T. Steyaert, P.L. Vidale, G.E. Liston, W.A. Lyons, and T.N. Chase, 1999: The influence of anthropogenic landscape changes on weather in south Florida. *Monthly Weather Review,* **127(7)**, 1663-1673.

Pinto, J.G., E.L. Frohlich, G.C. Leckebusch and U. Ulbrich, 2007: Changing European storm loss potential under modified climate conditions according to ensemble simulations of the ECHAM5/MPI-OM1 GCM. *Natural Hazards and Earth System Sciences,* **7(1)**, 165-175.

Pryor, S.C., R.J. Barthelmie, and J.T. Schoof, 2006: Interannual variability of wind indices across Europe. *Wind Energy,* **9**, 27-38.

Pulwarty, R., 2003: Climate and water in the West: science, information and decision-making. *Water Resources Update,* **124**, 4-12.

Saha, S., S. Nadiga, C. Thiaw, J. Wang, W. Wang, Q. Zhang, H.M. Van den Dool, H.-L. Pan, S. Moorthi, D. Behringer, D. Stokes, M. Peña, S. Lord, G. White, W. Ebisuzaki, P. Peng, and P. Xie, 2006: The NCEP climate forecast system. *Journal of Climate,* **19(15)**, 3483-3517.

Santer, B.D., K.E. Taylor, T.M.L. Wigley, T.C. Johns, P.D. Jones, D.J. Karoly, J.F.B. Mitchell, A.H. Oort, J.E. Penner, V. Ramaswamy, M.D. Schwarzkopf, R.J. Stouffer, and S. Tett, 1996: A search for human influences on the thermal structure in the atmosphere. *Nature,* **382(6586)**, 39-45.

Schubert, S., D. Dee, S. Uppala, J. Woollen, J. Bates, and S. Worley, 2006: *Report of the Workshop on "The Development of Improved Observational Data Sets for Reanalysis: Lessons Learned and Future Directions".* University of Maryland Conference Center, College Park, MD, 28-29 September 2005. [NASA Global Modeling and Assimilation Office, Greenbelt, MD], 31 pp. <http://gmao.gsfc.nasa.gov/pubs/docs/Schubert273.pdf>

Schwartz, M.N. and R.L. George, 1998: *On the Use of Reanalysis Data for Wind Resource Assessment.* NREL/CP-500-25610. National Renewable Energy Laboratory, Golden, CO, 5 pp. <http://www.nrel.gov/docs/fy99osti/25610.pdf>

Simmons, A., S. Uppala, and K.E. Trenberth, 2006: Future needs in atmospheric reanalysis. *EOS, Transactions of the American Geophysical Union,* **87(51)**, 583.

Stone, D.A. and M.R. Allen, 2005: The end-to-end attribution problem: from emissions to impacts. *Climatic Change,* **71(3)**, 303-318.

Stott, P.A., D.A. Stone, and M.R. Allen, 2004: Human contribution to the European heat wave of 2003. *Nature,* **432(7017)**, 610-614.

Trenberth, K., E.B. Moore, T.R. Karl, and C. Nobre, 2006: Monitoring and prediction of the Earth's climate: A future perspective. *Journal of Climate,* **19(20)**, 5001-5008.

Whitaker, J.S. and T. M. Hamill, 2002: Ensemble data assimilation without perturbed observations. *Monthly Weather Review,* **130(7)**, 1913-1924.

Whitaker, J.S., G.P. Compo, X. Wei, and T.M. Hamill, 2004: Reanalysis without radiosondes using ensemble data assimilation. *Monthly Weather Review,* **132(5)**, 1190-1200.

Zupanski, D. and M. Zupanski, 2006: Model error estimation employing an ensemble data assimilation approach. *Monthly Weather Review,* **134(5)**, 1337-1354.

APPENDIX B REFERENCES

Brohan, P., J.J. Kennedy, I. Harris, S.F.B. Tett, and P.D. Jones, 2006: Uncertainty estimates in regional and global observed temperature changes: a new dataset from 1850. *Journal of Geophysical Research,* **111**, D12106, doi:10.1029/2005JD006548.

Chen, M., P. Xie, J.E. Janowiak, and P.A. Arkin, 2002: Global land precipitation: a 50-year monthly analysis based on gauge observations. *Journal of Hydrometeorology,* **3(3)**, 249-266.

Hansen, J., R. Ruedy, M. Sato, M. Imhoff, W. Lawrence, D. Easterling, T. Peterson, and T. Karl, 2001: A closer look at United States and global surface temperature change. *Journal of Geophysical Research,* **106(D20)**, 23947-23963.

IPCC (Intergovernmental Panel on Climate Change) 2007a: *Climate Change 2007: The Physical Science Basis.* Contribution of Working Group I to the Fourth Assessment Report (AR4) of the Intergovernmental Panel on Climate Change [Solomon, S., D. Qin, M. Manning, Z. Chen, M. Marquis, K.B. Averyt, M. Tignor, and H.L. Miller (eds.)]. Cambridge University Press, Cambridge, UK, and New York, 996 pp. <http://www.ipcc.ch>

Kalnay, E., M. Kanamitsu, R. Kistler, W. Collins, D. Deaven, L. Gandin, M. Iredell, S. Saha, G. White, J. Woollen, Y. Zhu, A. Leetmaa, R. Reynolds, M. Chelliah, W. Ebisuzaki, W. Higgins, J. Janowiak, K.C. Mo, C. Ropelewski, J. Wang, R. Jenne, and D. Joseph, 1996: The NCEP/NCAR 40-Year Reanalysis Project. *Bulletin of the American Meteorological Society,* **77(3)**, 437-471.

NCDC (National Climatic Data Center), 1994: *Time Bias Corrected Divisional Temperature-Precipitation-Drought Index.* Documentation for dataset TD-9640. National Climatic Data Center, Asheville, NC, 12 pp. <http://www1.ncdc.noaa.gov/pub/data/documentlibrary/tddoc/td9640.pdf>

Peterson, T.C., T.R. Karl, P.F. Jamason, R. Knight, and D.R. Easterling, 1998: First difference method: Maximizing station density for the calculation of long-term global temperature change, *Journal of Geophysical Research,* **103(D20)**, 25967-25974.

Roeckner, E., K. Arpe, L. Bengtsson, M. Christoph, M. Claussen, L. Dümenil, M. Esch, M. Giorgetta, U. Schlese, and U. Schulzweida, 1996: *The Atmospheric General Circulation Model ECHAM-4: Model Description and Simulation of Present-Day Climate.* MPIM Report 218. Max-Planck-Institute for Meteorology, Hamburg, Germany, 90 pp.

Rudolf, B. and U. Schneider, 2005: Calculation of gridded precipitation data for the global land-surface using in-situ gauge observations. In: *Proceedings of the 2nd Workshop of the International Precipitation Working Group (IPWG),* Monterey, October 2004, pp. 231-247.

Santer, B.D., W. Brüggemann, U. Cubasch, K. Hasselmann, E. Maier-Reimer, and U. Mikolajewicz, 1994: Signal-to-noise analysis of time-dependent greenhouse warming experiments. Part 1: pattern analysis. *Climate Dynamics,* **9(6)**, 267-285.

Schubert, S.D., M.J. Suarez, P.J. Pegion, R.D. Koster, and J.T. Bacmeister, 2004: Causes of long-term drought in the U.S. Great Plains. *Journal of Climate,* **17(3)**, 485-503.

Smith, T.M. and R.W. Reynolds, 2005: A global merged land-air-sea surface temperature reconstruction based on historical observations (1880-1997). *Journal of Climate,* **18(12)**, 2021-2036.

Von Storch, H. and F.W. Zwiers, 1999: *Statistical Analysis in Climate Research.* Cambridge University Press, Cambridge, UK, and New York, 484 pp.

PHOTOGRAPHY CREDITS

Cover/Title Page/Table of Contents
Image for Chapter 3, page 47, (Doppler), NOOA photo
Image for Chapter 4, page 93, (Polar researcher), NOAA, Micahel Van Woert

Chapter 2
Page 28, (Polar stratospheric clouds), Mark Schoeberl, GSFC/NASA
Page 39, (Weather balloon launch), Mike Dunn, NOAA/CPO

Chapter 4
Page 93, (Polar research team), NOAA/POPS
Page 95, (Research ship workers recovering equipment), NOAA
Page 99, (Polar researchers), NOAA Office of Ocean Exploration, J. Potter
Page 103, (Setting up equipment), NOAA/CPO, Mike Dunn

Contact Information

Global Change Research Information Office
c/o Climate Change Science Program Office
1717 Pennsylvania Avenue, NW
Suite 250
Washington, DC 20006
202-223-6262 (voice)
202-223-3065 (fax)

The Climate Change Science Program incorporates the U.S. Global Change Research Program and the Climate Change Research Initiative.

To obtain a copy of this document, place an order at the Global Change Research Information Office (GCRIO) web site: http://www.gcrio.org/orders

---

Climate Change Science Program and the Subcommittee on Global Change Research

William Brennan, Chair
Department of Commerce
National Oceanic and Atmospheric Administration
Director, Climate Change Science Program

Jack Kaye, Vice Chair
National Aeronautics and Space Administration

Allen Dearry
Department of Health and Human Services

Anna Palmisano
Department of Energy

Mary Glackin
National Oceanic and Atmospheric Administration

Patricia Gruber
Department of Defense

William Hohenstein
Department of Agriculture

Linda Lawson
Department of Transportation

Mark Myers
U.S. Geological Survey

Timothy Killeen
National Science Foundation

Patrick Neale
Smithsonian Institution

Jacqueline Schafer
U.S. Agency for International Development

Joel Scheraga
Environmental Protection Agency

Harlan Watson
Department of State

EXECUTIVE OFFICE AND OTHER LIAISONS

Robert Marlay
Climate Change Technology Program

Katharine Gebbie
National Institute of Standards & Technology

Stuart Levenbach
Office of Management and Budget

Margaret McCalla
Office of the Federal Coordinator for Meteorology

Rob Rainey
Council on Environmental Quality

Daniel Walker
Office of Science and Technology Policy